Air Pollution Control, Part II

ENVIRONMENTAL SCIENCE AND TECHNOLOGY

A Wiley-Interscience Series of Texts and Monographs

Edited by ROBERT L. METCALF, *University of Illinois*

JAMES N. PITTS, Jr., *University of California*

AIR POLLUTION CONTROL

Part II

EDITED BY

WERNER STRAUSS

University of Melbourne
Victoria, Australia

Wiley-Interscience

A DIVISION OF JOHN WILEY & SONS, INC.

NEW YORK · LONDON · SYDNEY · TORONTO

The paper used in this book has pH of 6.5
or higher. It has been used because the
best information now available indicates
that this will contribute to its longevity.

Library of Congress Catalogue Card Number: 69-18013

ISBN 0 471 83319 3

Printed in the United States of America

10 9 8 7 6 5 4 3 2 1

SERIES PREFACE

Environmental Science and Technology

The Environmental Science and Technology Series of Monographs, Textbooks, and Advances is devoted to the study of the quality of the environment and to the technology of its conservation. Environmental science therefore relates to the chemical, physical, and biological changes in the environment through contamination or modification, to the physical nature and biological behavior of air, water, soil, food, and waste as they are affected by man's agricultural, industrial, and social activities, and to the application of science and technology to the control and improvement of environmental quality.

The deterioration of environmental quality, which began when man first collected into villages and utilized fire, has existed as a serious problem under the ever-increasing impacts of exponentially increasing population and of industrializing society, environmental contamination of air, water, soil, and food has become a threat to the continued existence of many plant and animal communities of the ecosystem and may ultimately threaten the very survival of the human race.

It seems clear that if we are to preserve for future generations some semblance of the biological order of the world of the past and hope to improve on the deteriorating standards of urban public health environmental science and technology must quickly come to play a dominant role in designing our social and industrial structure for tomorrow. Scientifically rigorous criteria of environmental quality must be developed. Based in part on these criteria, realistic standards must be established and our technological progress must be tailored to meet them. It is obvious that civilization will continue to require increasing amounts of fuel, transportation, industrial chemicals, fertilizers, pesticides, and countless other products and that it will continue to produce waste products of all descriptions. What is urgently needed is a total systems approach to modern civilization through which the

v

pooled talents of scientists and engineers, in cooperation with social scientists and the medical profession, can be focused on the development of order and equilibrium to the presently disparate segments of the human environment. Most of the skills and tools that are needed are already in existence. Surely a technology that has created such manifold environmental problems is also capable of solving them. It is our hope that this Series in Environmental Sciences and Technology will not only serve to make this challenge more explicit to the established professional but that it also will help to stimulate the student toward the career opportunities in this vital area.

Robert L. Metcalf
James N. Pitts, Jr.

PREFACE TO PART II

The control of air pollution involves not only the technical ability to capture particles and gases before their emission to the atmosphere, but also informed decisions as to the extent to which this must be effected, the permissible levels of these emissions which seem to be an inescapable concomitant of so many of man's activities. Furthermore, in making these decisions we must take into account both the immediate and known local effects of pollutants and their predicted long-term effects as part of a worldwide situation. The latter is exceptionally difficult, and the outstanding work of Drs. Robinson and Robbins in the evaluation of total emissions, atmospheric conditions and global dispersion of air pollutants, as well as their eventual fate, constitutes the first chapter of the second volume.

The enforcement of decisions at governmental level about air pollution control—and here we usually mean source control—requires legislation, and the enforcement of legislation. Mrs. Lanteri, of the Melbourne University Law School has been studying such legislation in recent years, because it represents a departure from traditional concepts of "right and wrong," being concerned rather to legislate for the technical and economically feasible or practical. It is my hope, in a future volume, to include a chapter on the problems of grass roots enforcement.

An authoritative chapter on control of radioactive airborne wastes from nuclear reactors is also timely, as numerous orders for these reactors have been placed in the past 3 years for the 1970's, which will see nuclear power generation as a substantial contributor to our power supplies.

This volume also discusses the possibility of using thermal forces as a collection mechanism for particulate collectors. While this method has received intermittent attention, as have sonic agglomeration methods, no commercial equipment has resulted. The literature of the subject is, of course, the basic material needed by all working in the field of air pollution, and the extensive list of source material will be a considerable help to the increasing numbers of scientists, technologists and engineers who are developing and applying air pollution control technology.

Subsequent volumes, which are scheduled to appear at intervals of 18 months, will further develop these themes.

Melbourne, Australia   WERNER STRAUSS

May, 1971

PREFACE TO PART I

The increasing civilization of man unfortunately brings with it increasing pollution of his environment. In recent years we have become aware of this pollution, and studies of the control of solid, liquid, and gaseous wastes are being undertaken by a large number of universities, research institutes, and industrial research organizations. This has led to a rapid growth in the literature on pollution and its control, and it is becoming difficult to remain aware of current work in some of the critical areas.

The main aim of this volume on air pollution control is to present authoritative reviews of specific fields currently of major importance. With this in mind, the authors of the individual chapters have been selected because of their current active participation in the subject.

The reviews attempt to be not only comprehensive but also selective and critical in the constructive sense. Whenever possible, worked examples that illustrate methods of calculation have been included.

It is my hope that the papers will provide an adequate background in their specific fields for engineers and scientists who are faced with particular air pollution control problems. Furthermore, they should be helpful to those who are working in the field application of processes, methods, or techniques as well as to those who are engaged in research for better control methods.

I should like to thank the authors for their cooperation in following the rather stringent guidelines laid down, which have resulted in the patterns of the different chapters aimed at presenting the critical "state of the art." I also wish to express my appreciation for the help I received from my colleagues at the University of Melbourne, Australia, and at the Westinghouse Research and Development Center, Pittsburgh, Pennsylvania—in particular, Dr. J. Bagg, Dr. E. V. Somers, Dr. B. W. Lancaster, and Mr. J. R. Hamm. Much of the typing of the manuscript and the compilation of tables and diagrams was done by Mrs. Tracee Tang, without whose assistance this volume would not have been produced so speedily and efficiently and to whom I wish to express my gratitude.

Melbourne, Australia   WERNER STRAUSS
March 1970

CONTENTS

Air Pollution Control, Part II

Emissions, Concentrations, and Fate of Gaseous Atmospheric Pollutants

Elmer Robinson and Robert C. Robbins

Stanford Research Institute, Menlo Park, California

TABLES

I. INTRODUCTION

It is a popular misconception that almost everything present in the earth's atmosphere, except nitrogen, oxygen, water, and a few inert rare gases, are air pollutants or the remnants of pollution activities. This is not the case. There are sources in the natural environment of a wide variety of gaseous and particulate materials which are commonly classed as air pollutants when they are emitted by man-made sources. In addition, scavenging mechanisms exist in the natural environment for gaseous pollutants and these have a direct bearing on the probability of long-term accumulations of pollutants in the atmosphere. However, it should be borne in mind that these scavenging mechanisms do not operate fast enough to provide solutions for local urban air pollution situations. In fact, the existence of an air pollution problem is evidence that the available scavenging mechanisms have been overburdened by the local pollutant emission rates.

In discussions of air pollution, the assumption has often seemed to be tacitly made that pollutants once released into the atmosphere will disappear as soon as the air mass moves across the city boundary. However, just as often, it has also probably been assumed that commonly identified pollutant were uniquely related to man's activity, and that atmospheric pollutant emissions constitute a permanent and ever-worsening burden for our atmospheric environment. Both of these assumptions are drastic over-simplifications and both can lead their proponents far from reality.

It is the purpose of this review to examine our present state of knowledge of the atmospheric cycles followed by a variety of common gaseous materials. This will be done by analyzing the major natural and urban pollution sources, estimating the effectiveness of applicable atmospheric reaction processes, and then determining the effectiveness of the processes by which the material is finally removed from the atmosphere. It will be apparent that our knowledge of these atmospheric processes is, in many cases, fragmentary and based on rather unsatisfactory assumptions. However, a considerable amount of research is presently being done to expand our information.

Sources of the various materials have been estimated from the available literature, although a detailed intercomparison of source data has not been undertaken. In general, the magnitudes of the various pollutant sources can be estimated reasonably well while the possible natural sources can only be crudely defined.

We have, in a general sense, divorced our considerations from a detailed reporting and analysis of conditions within specific urban areas and from those data which report the periodic "highest-ever" type of results. The total atmospheric system is so much greater than the polluted envelope of any one city that the transient effects noted in a given area are important to our study only to the degree that they indicate sources of pollutants that can affect the atmospheric system.

II. ATMOSPHERIC SULFUR COMPOUNDS

A. Introduction

The emission of sulfur compounds into the atmosphere has developed into one of the major pollution problems of the late 1960's. The interest in air pollution is centered almost exclusively on the emission of SO_2, although some attention has been given to SO_3 which is present in most SO_2 emission streams (1).

In the following discussion of sulfur compounds in the atmosphere three forms of atmospheric sulfur—SO_2, H_2S, and sulfates (SO_4)*—will be considered. The emphasis will be on the atmospheric phases of the cycle of the various sulfur compounds through the environment. This analysis will consider the forms and quantities of sulfur which are emitted into the atmosphere as well as the concentrations of the compounds occurring in the ambient atmosphere in order to present an adequate discussion. It will not be necessary to consider either emissions or atmospheric concentrations in great detail or sophistication. Thus, most of the data used will be averages from fairly general data sources. Details are, of course, available to the reader in many reference reports and studies.

First, we will examine the sources for the atmospheric sulfur compounds SO_2, H_2S, and SO_4, and then, their various atmospheric reactions. The

---

* For convenience the notation SO_4 will be used in this presentation with the understanding that this is the sulfate ion and that it is actually present as a sulfate compound such as $(NH_4)_2SO_4$ or H_2SO_4.

discussion is concluded with analyses of current atmospheric concentrations and the cycle of sulfur through the environment.

We have not considered SO_3 separately.   It is obviously the product of the gas phase oxidation of SO_2, although there is no evidence that SO_3 is present under normal atmospheric conditions since it so quickly converts to H_2SO_4 and thus is included under SO_4.   In some references SO_3 is indicated as a minor industrial emission along with SO_2, but it has usually not been considered to be a separate emission.

B. Sources of Atmospheric Sulfur Compounds

The sulfur in the atmosphere comes from both the natural environment and air pollution emissions.   The natural sulfur is represented by sulfate aerosols produced in sea spray, as well as by H_2S from the decomposition of organic matter in swamp areas, bogs, and tidal flats.   Volcanic areas are also sources of natural H_2S.   Some H_2S also reaches the atmosphere as a result of industrial operations and from a variety of pollution sources.   The SO_2 which is released to the atmosphere comes almost exclusively from pollution sources.   The only natural SO_2 source which is normally mentioned is volcanic activity.

1. Sources of Sulfates

An estimate of the sea spray production of SO_4 has been made by Eriksson (2).   His estimate of the annual production of SO_4 from sea spray is 130 × 10^6 tons.   However, of this only 10%, or 13 × 10^6 tons penetrates continental areas, while the remaining 90% is precipitated over the oceans Junge (3) also discounts the significance of sea spray as a source of continental SO_4 aerosols and considers that spray production and precipitation is essentially a closed cycle within the marine environment.

Sea spray is only one of three possible sources for SO_4 aerosols found in the atmosphere. The other two sources are the oxidation of SO_2 and H_2S to form H_2SO_4 or SO_4 salts.   These mechanisms will be discussed later.   On the basis of a study of sulfur isotope ratios, Jensen and Nakai (4) concluded that in industrial areas the atmospheric SO_4 was due primarily to oxidized SO_2 and in non-industrial areas the oxidation of H_2S provided the observed SO_4.   Although they also stated that the SO_4 found in precipitation is not likely to originate in sea spray, the available data are not sufficient for a clear delineation of the relative effectiveness of these several processes as sources for the atmospheric SO_4.   The fact that these data are based on analyses of SO_4 and S isotope ratios in rain complicates the extrapolation to the ambient atmospheric aerosol because of different collection efficiencies of rain and cloud drops for gaseous and particulate sulfur compounds.

However, Östlund (5) also concluded, after a study of isotopes, that SO_4 in precipitation comes from the oxidation of gaseous compounds rather than sea spray.

2. Hydrogen Sulfide Sources

The magnitude of H_2S production from swamps and other areas of decaying organic matter have only been speculative because there are no data upon which to base any estimates of the concentration of H_2S in the ambient atmosphere. Eriksson (2) concluded that the annual H_2S production from the oceans was 202×10^6 tons (190×10^6 tons S) on the basis of the amount of S needed to balance his environmental sulfur cycle. He also estimated that decaying vegetation on land areas would produce 82×10^6 tons of H_2S (77×10^6 tons S). This was based on land area productivity data and an average vegetation sulfur content of 0.5 mg S per gram of CO_2 stored in the organic matter. Thus Eriksson's estimated total annual H_2S production is 284×10^6 tons H_2S (267×10^6 tons S). Junge (3) arrived at essentially the same estimates for ocean and land areas. He also points out that a transfer of 30×10^6 tons of sulfur from ocean to land in the form of H_2S is indicated by his calculations. In later discussions Junge (6) states that such a transfer could only result from a situation where S concentrations over the ocean were larger than over land areas and that this is contrary to all observations.

Research by Östlund and Alexander (7) led to the conclusion that it is very unlikely that significant amounts of H_2S could escape from the sea surface because the solubility and reactivity of the gas causes its oxidation and retention in the ocean. The only exceptions might be limited areas of oxygen-deficient water. In our own analysis and as a result of a new calculation of the sulfur cycle, we estimate an annual ocean area H_2S emission of 30×10^6 tons and a land area emission of about 70×10^6 tons. Thus, we conclude on the basis of circumstantial evidence that ocean areas are apparently sources of H_2S. In addition, there is good reason to suspect that some emissions of H_2S could occur in various localized marine areas. The estimated release of between 60 and 80×10^6 tons of H_2S over land areas from decaying vegetation seems reasonable.

Various industrial operations, notably kraft paper mills and oil refining, can emit H_2S as part of their atmospheric emanations. These emissions, however are seldom catalogued. Because there are difficulties in analyzing H_2S in the presence of SO_2, it is frequently lumped in with SO_2. The fact that serious odor problems can result from even minor volume emissions of H_2S means that industrial emissions will be minimized wherever possible. A rough estimate for kraft paper mill operations indicates that the world-wide production of about 40×10^6 tons of kraft pulp in the mid-1960's is linked with an emission of 0.06×10^6 tons of H_2S. A study of Jacksonville, Florida, air pollution (8) indicates that industrial H_2S may amount to about 2% of the SO_2 emitted. This factor, if applied to a world-wide SO_2 emission of 146×10^6 tons, would indicate an industrial H_2S emission of about 3×10^6 tons. No more comprehensive or quantitative estimates of industrial H_2S emissions have apparently been made, but the industrial tonnage of H_2S is insignificant in the total sulfur cycle.

3. Sulfur Dioxide Sources

The SO_2 emissions to the atmosphere from pollution sources have been estimated on a world-wide basis from data which are generally applicable to the mid-1960's. The results of this estimate are listed in Table I and show a total annual SO_2 emission of 146×10^6 tons (73×10^6 tons S). Of this total, 70% is estimated to come from coal combustion and 16% from the combustion of petroleum products, mainly residual fuel oil. The remaining tonnage is accounted for by petroleum refining operations, about 6×10^6 tons, and by non-ferrous smelting, about 14×10^6 tons.

These estimates were based primarily on 1965 world figures for coal production, petroleum refining activity and products, smelter production

TABLE I
Worldwide Annual Emissions of Sulfur Dioxide

Source	Product consumption or production rate	SO_2 Production factor	SO_2 Emission, tons
Coal[a]	$3,074 \times 10^6$ T[b]	3.3 T/100 T[c]	102×10^6
Petroleum combustion			
Gasoline	379×10^6 T[b]	9×10^{-4} T/T[c]	0.3×10^6
Kerosene	100×10^6 T[b]	24×10^{-4} T/T[c]	0.2×10^6
Distillate	287×10^6 T[b]	70×10^{-4} T/T[c]	2.0×10^6
Residual	507×10^6 T[b]	400×10^{-4} T/T[c]	20.3×10^6
Petroleum refining	$11,317 \times 10^6$ bbl[b]	50 T/10^5 bbl[c]	5.7×10^6
Smelting			
Copper	6.45×10^6 T[d]	2.0 T/T[e]	12.9×10^6
Lead	3.0×10^6 T[d]	0.5 T/T[e]	1.5×10^6
Zinc	4.4×10^6 T[d]	0.3 T/T[e]	1.3×10^6
			146.2×10^6 tons
			$(1.32 \times 10^{14}$ g)

[a] Includes lignite.

[b] U.S. Statistical Abstracts, Table No. 1256, 1967 (93).

[c] Derived from Rohrman and Ludwig 1965 (9).

[d] U.S. Bureau of Mines, Mineral Trade Notes 1967 (11).

[e] Stanford Research Institute derived from 1967 smelter data.

combined with SO_2 emission factors per unit of production derived from U.S. industrial operations by Rohrman and Ludwig (9), and by Stanford Research Institute (10).

Since coal combustion is the major source of SO_2, some further discussion of this source is justified. The estimated emission of 102×10^6 tons is dependent on the conversion factor of 3.3 tons of SO_2 per 100 tons of coal produced. This factor comes from Rohrman and Ludwig (9) who have estimated a United States production of 14×10^6 tons of SO_2 from a coal production of 425×10^6 tons. They point out that although United States coal averages 2% sulfur, indicating an emission factor of 4T/100T, about 10% of the S in coal remains in the ash. In addition, the final emission factor is also altered by the shrinkage in transit of the tabulated mine production and by the diversion of some coal to uses which do not release sulfur products. Extending these U.S. data to world coal emissions is admittedly imprecise; however, it seems likely that the estimates are within $\pm 20\%$ of the correct value.

Previous estimates of world sulfur emissions were made by Katz (12) who provided data covering the period 1937–1947. For the years of 1937 and 1940, total sulfur emissions (as SO_2) were 69×10^6 tons and 78×10^6 tons, respectively. These data of Katz have been used by both Junge (3) and Eriksson (2) in their atmospheric sulfur cycle calculations. Thus, emissions have roughly doubled in the period between 1940 and 1965.

4. General Emissions

These source data for SO_4, H_2S, and SO_2 have been combined in Table II to show the estimated world natural and pollution emissions of sulfur compounds. The total emission, expressed as sulfur, is 220×10^6 tons. This

TABLE II

Annual Amounts of Sulfur Introduced into the Atmosphere

Emission	Source	Amount, tons	Amount as sulfur, tons
SO_4 Aerosol	Sea spray	130×10^6 [a]	44×10^6
H_2S	Biological decay in ocean	30×10^6 [b,c]	30×10^6
	Biological decay on land	60–80×10^6 [a,b,d]	70×10^6
	Pollutants	3×10^6 [b]	3×10^6
SO_2	Pollutants	146×10^6 [b]	73×10^6
			220×10^6

[a] Eriksson 1959, 1960 (2).

[b] This study.

[c] Eriksson estimates this as high as 200×10^6 tons H_2S.

[d] Junge 1963 (3).

estimate is not very precise because about 50% of this total comes from estimates of H_2S emissions from land and ocean areas; and as previously mentioned, there is little or no observational data to support an H_2S estimate. It is interesting, however, to note that industrial emissions may account for only 35% of the sulfur released annually to the atmosphere. Even considering land areas alone, industrial emissions account for only about half of the total annual sulfur.

5. Hemispheric Sulfur Dioxide Emissions

Table III shows the total SO_2 emissions divided according to sources into Northern and Southern hemispheres. On a total basis, 93% of the SO_2 emitted by pollutant sources is emitted in the Northern hemisphere. The total is estimated at 136×10^6 tons from Northern hemispheric sources out of a total 146×10^6 tons.

Table IV shows our estimate of the total hemispheric emissions of sulfur in its various emission forms—SO_2, H_2S, and sulfate. Here H_2S emissions have been prorated over the warmer land and ocean, those areas between 65° S and 65° N, while sea spray sulfates have been prorated according to total ocean areas. This tabulation shows that the atmosphere of the Northern hemisphere receives over twice as much sulfur as does the Southern hemisphere. The ratio is 149×10^6 tons or 69% of the global total in the Northern hemisphere compared with 66×10^6 tons or 31% in the Southern hemisphere.

The hemispheric imbalance in SO_2 emissions points out the opportunity that exists to determine the effects, if any, that SO_2 pollutant emissions may exert on atmospheric parameters. Parallel experiments in the Northern

TABLE III
Hemispheric SO_2 Pollutant Emissions
(10^6 tons)

Source	Total SO_2	Northern hemisphere		Southern hemisphere	
Coal[a]	102	98	(96%)	4	(4%)
Petroleum[a]					
Comb. & Refin.	28.5	27.1	(95%)	1.4	(5%)
Smelting[b]					
Copper	12.9	8.6	(67%)	4.3	(33%)
Lead	1.5	1.2	(80%)	0.3	(20%)
Zinc	1.3	1.2	(90%)	0.1	(10%)
TOTAL	146	136	(93%)	10	(7%)

[a] United Nations Statistical Papers—World Energy Supplies 1963–1966, Series J, No. 11 Tables 2 and 9 (13).

[b] U.S. Bureau of Mines, Mineral Trade Notes Vol. 64, No. 9 and 12 1967 (11).

TABLE IV

Total Hemispheric Sulfur Emissions

(10^6 tons S)

Source	Total	Northern hemisphere	Southern hemisphere
Pollutant SO_2 sources	73	68	5
Biological H_2S (land)	68	49[a]	19[a]
Biological H_2S (marine)	30	13[b]	17[b]
Sea spray	44	19[c]	25[c]
TOTAL	215	149 (69%)	66 (31%)

[a] Based on ratio of land area between 0 and 65° N and S.

[b] Based on ratio of ocean areas between 0 and 65° N and S.

[c] Based on ratio of ocean areas in both hemispheres.

and Southern hemispheres should provide a situation where most factors except pollutant backgrounds could be generally equated and thus, the impact of the pollutants on the atmosphere could be evaluated.

6. Past and Future Sulfur Dioxide Emissions

Sulfur dioxide has been a major pollutant, at least since the first smelting of copper sulfide ore and the beginning of the widespread burning of soft coal, throughout the history of industrial processes.

Table V shows the SO_2 pollutant emissions for the period 1860–1965 as estimated from available fuel and smelting data.   Figure 1 shows the total SO_2 estimate as a function of time over this 100-year period.

Figure 1 clearly shows that there has been a major increase of about 65×10^6 tons in SO_2 emissions between 1950 and 1965.   As shown in Table V 36×10^6 tons or 55% of this increase is related to increased coal combustion and 20×10^6 tons or 31% is due to increased petroleum usage (13).   The effects of the depression in the 1930's and World War II in the 1940's are evident in these data, although the 10-year spacing of data points does not emphasize these factors.   The lack of readily available production data for lead and zinc before 1900 is not serious in judging the trend in SO_2 emissions.

In the 25 years between 1940 and 1965, SO_2 emissions have essentially doubled.   Before 1940 it had taken 35 years, since 1905, for SO_2 emissions to double.   Other estimates have been made of global SO_2 emissions; Katz (12), for example, estimated 1940 SO_2 emissions at 78×10^6 tons as compared with our 72×10^6 tons.   Agreement within 10% is certainly satisfactory when the approximations entering into these estimates are considered.

TABLE V
Estimated Historical SO$_2$ Emissions*
(10^6 tons)

	Coal[a]	Petroleum[b]	Copper[c]	Lead[d]	Zinc[e]	Total
1860	5.0	0.0	0.22			5.22
1870	7.8	0.01	0.24			8.05
1880	12.2	0.07	0.26			12.53
1890	18.7	0.17	0.68			19.55
1900	28.1	0.33	1.60	0.47	0.15	30.65
1910	42.1	0.70	2.84	0.61	0.26	46.51
1920	49.5	1.79	3.26	0.54	0.22	55.31
1930	51.3	3.12	3.52	0.93	0.46	59.33
1940	61.0	4.62	5.46	0.97	0.53	72.58
1950	66.0	8.30	5.92	0.86	0.60	81.68
1960	95.7	19.9	10.0	1.28	1.0	127.88
1965	102.0	28.5	12.9	1.5	1.3	146.20

* Based on the following production data references:

[a] Peaceful Uses of Nuclear Energy, UN, 1956, Table XXIII (58).

[b] Peaceful Uses of Nuclear Energy, UN, 1956, Table XXIII (58).

[c] McMahon, A. D., Copper—A Materials Survey, U.S. Bureau of Mines I.C. 8225, (1965), Table 26 (14); and Herfindahl, O. C., Copper Costs and Prices: 1870–1957, Johns Hopkins Press, 1959, Tables 2 and A-1 (15).

[d] U.S. Bureau of Mines, Materials Survey: Lead, 1951 (16); and UN International Lead and Zinc Study Group, Factors Affecting Consumption, Appendix 1 1966 (17).

[e] U.S. Bureau of Mines, Materials Survey: Zinc, 1951 (18); and UN International Lead and Zinc Study Group, Factors Affecting Consumption, Appendix 1, 1966 (17).

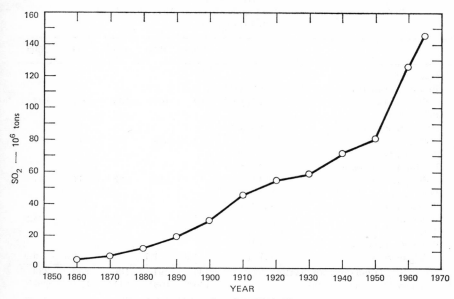

Fig. 1. Estimated historical SO$_2$ emissions.

11

The emissions of SO_2 have changed rapidly in the past 10 years and further increases can be expected, although limits on sulfur content in fuels may be significant factors in any future projections. Table VI presents our estimate of global SO_2 emissions from now until the year 2000 if no allowances are made for the introduction of additional emission controls. Figure 2 shows the total estimated global SO_2 emissions for the period 1940 to 2000.

As indicated in Table VI, coal consumption is expected to increase at the smallest rate until the year 2000, i.e., 0.2% per year (19). The rest of the possible SO_2 sources are expected to increase at appreciably faster rates, petroleum at 6.2% per year (19), copper smelting at 4.3% per year (17), lead smelting at 6.3% per year (17), and zinc smelting at 6.1% per year (17). If SO_2 emissions also increase at these same rates, by the year 2000, estimated SO_2 emissions would be 333×10^6 tons annually. In this projection it will be after 1990 before coal ceases to be the largest single source of SO_2.

These projected emissions can also be related to our estimate of natural emissions. Tables II and IV indicate that natural sources annually account for about 144×10^6 tons of sulfur on a global basis and about 81×10^6 tons in the Northern hemisphere. According to our SO_2 emission projections, global pollutant emissions will exceed natural sources of sulfur after about 1995. In the Northern hemisphere, if 90% of the SO_2 is attributed to

TABLE VI

Projected SO_2 Emissions 1965–2000 Based on 1965 Emission Factors
(10^6 tons/year)

		1965	1970	1980	1990	2000
Coal						
Production increase	0.2% yr^{-1}					
SO_2 increase	0.2% yr^{-1}	102	103	105	107	109
Petroleum						
Production increase	6.2% yr^{-1}					
SO_2 increase	6.2%	29	38	62	100	162
Smelting copper						
Production increase	4.3% yr^{-1}					
SO_2 increase	4.3%	13	16	23	33	47
Lead						
Production increase	6.3% yr^{-1}					
SO_2 increase	6.3	1.5	2.0	3.3	5.4	8.8
Zinc						
Production increase	6.1% yr^{-1}					
SO_2 increase	6.1%	1.3	1.7	2.7	4.3	6.9
TOTAL SO_2 (1965 CONTROL RATE)		147	161	196	250	333
PERCENT RELATIVE TO 1965		100%	109%	133%	170%	226%

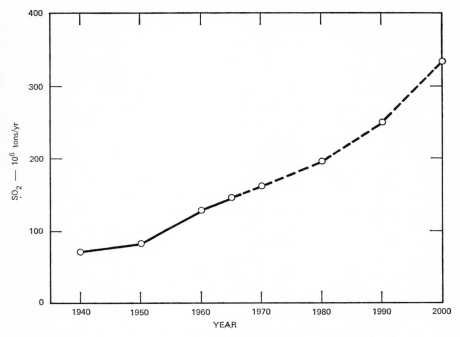

Fig. 2. Projected global SO_2 emissions based on 1965 emission factors.

Northern hemispheric sources, pollutant sources of sulfur will exceed the natural sources by about 1975. New control measures on major sulfur emissions can be expected to affect these estimates, however.

C. Atmospheric Reactions of Sulfur Compounds

1. Sulfur Dioxide Reactions in the Atmosphere

In contrast to many pollutant gases in the atmosphere, sulfur dioxide is both very soluble in water, 11.28 g/100 ml at 20° C, and very reactive in the atmosphere in dilute concentrations either photochemically or catalytically. The end product of these SO_2 reactions is H_2SO_4 or a sulfate salt.

The work on SO_2 atmospheric reactions which seems to be most frequently quoted is that done in the 1950's in the laboratory of Dr. H. F. Johnstone at the University of Illinois. The reaction or oxidation rate of SO_2 in small H_2SO_4 droplets was studied by Gerhard and Johnstone (20), and they found that in the absence of a catalyst the oxidation rate of SO_2 was negligible. They also estimated that the oxidation of SO_2 in sunlight was about 0.1 to 0.2 % per hour independent of NO_2, NaCl, or relative humidity between 30 %

and 90%. Renzetti and Doyle (21) showed that there was appreciable formation of H_2SO_4 aerosol by 3130 Å radiation from SO_2 at concentrations of less than 1 ppm and humidity less than 50%. Their oxidation rates greatly exceeded those of Gerhard and Johnstone (21); however, Renzetti did not claim that their rates could be directly applied to the atmosphere. More recently, the 1955 studies of Gerhard and Johnstone have been challenged by Junge and Ryan (22) who, with very carefully controlled experiments, found very much lower rates of SO_2 photo-oxidation than were found in the 1955 experiments.

Interest in the rate of SO_2 oxidation in small water droplets evolved from analyses of several major air pollution events such as Donora, Pennsylvania, and London in which heavy fog occurred concurrently with serious air pollution effects and high concentrations of SO_2. However, it was characteristic for SO_2 concentrations to rise rather rapidly and then to remain more or less stable for the remainder of the event. This lead to the postulation of removal rates linked to the fog and the concentration of SO_2. The effects of various catalysts to promote SO_2 oxidation in fog droplets were studied by Junge and Ryan (22) and by Johnstone and Coughanowr (23). The Junge studies tested iron chloride as one possible catalyst but determined that manganese salts were more effective as was also found by Johnstone. Johnstone and Coughanowr determined that manganese sulfate in fog droplets would catalyze SO_2 oxidation at a rate of about 1% per minute.

In another examination of this point Gall (24) investigated the oxidation rate of SO_2 in the liquid phase in the presence of catalytic metal ions like Mg^{2+}. For conditions of London air (300 μg SO_2/m^3) and reasonable values for the Mg^{2+} concentrations, he estimates a decrease of SO_2 of 1% to 2% per hour corresponding to life times of about three days. In this case, it can be shown that the rate of oxidation within the droplets is the controlling factor.

Even with this latest work the presence of the catalyzing salts in fog droplets is frequently challenged with regard to the practicality of the system to actually explain the oxidation of SO_2 in polluted fogs. Doubts arise because there has never been any significant evidence that these specific materials occurred in the proper catalytic form or concentration in foggy atmospheres. Probably a more realistic system incorporating ammonia, also an emission from coal combustion along with SO_2, was studied both by Junge and Ryan (22) and by van den Heuvel and Mason (25). It was found by Junge and Ryan that because SO_2 had low solubility in water at low pH, ammonia promoted the solubility of SO_2 in droplets by neutralizing the acid formed in the droplets. The importance of the oxidation of SO_2 in the presence of NH_3 and fog droplets to form $(NH_4)_2SO_4$ was confirmed by van den Heuvel and Mason (25). They found the production rate to be roughly proportional to the product of exposure time and the square of the cloud droplet radius.   If diffusion of the gases toward the droplet were the rate controlling factor the rate should be proportional to r.   The observations thus point to the diffusion within the liquid phase as the rate controlling factor. Extrapolating these results from the laboratory to natural conditions gives an estimate that a cloud of droplets of 10 μ radius in an atmosphere of 100 μg SO_2/m^3 could produce $10^{-11}\,g$ $(NH_4)_2SO_4$ per hour per drop, resulting in an SO_2 lifetime of only about one hour.   This appears extremely short but indicates the efficiency of the process under the optimum conditions.

The research studies show that oxidation of SO_2 to SO_4 can proceed at a very rapid rate where there are liquid droplets present.   While it is certainly true that most SO_2 is not emitted under foggy conditions this type of emission has specific application to a large amount of power plant SO_2 because of the water vapor in the effluent and the droplet formation by condensation within the plume.   Gartrell et al. (26) studied the SO_2/SO_4 ratio in the Tennessee Valley Authority power plant plumes using helicopter transects at successive distances downwind.   They found very rapid, from about 0.1 up to 2 % per minute, oxidation rates.   In addition there was a direct relationship to ambient relative humidity and the presence of liquid water droplets within the plume.   Gartrell concluded that within the specific environment provided by the plume from a coal burning power plant SO_2 oxidation could be very rapid and that this rate was a function of the moisture present in the plume, either as an effluent component or from the ambient atmosphere.

After extensive studies of washout and rainout in field measurements Georgii (27) found indications that SO_2 may be removed directly by precipitation.   Subsequently he conducted a detailed experimental study of this process and confirmed the initial observations (28).   Their estimates for the removal rates under natural conditions on the basis of these experiments, show that the SO_4 content in rain over polluted areas at least may be largely due to direct washout of SO_2 encountered by the precipitation falling out of the clouds.

Sulfur dioxide oxidation is not confined to fog or to other conditions accompanied by liquid droplets.   While direct oxidation by molecular oxygen has been shown to be insignificant, photochemical oxidation of SO_2 in the presence of NO_2 and hydrocarbons to form aerosols is probably one of the more significant reaction systems for SO_2 in polluted atmospheres.   A relatively large number of studies of this system have been made.   Those carried out by Doyle and his co-workers on several Stanford Research Institute projects seem to be most directly applicable to a basic evaluation of the fate of SO_2 in the ambient atmosphere (21, 29).   These experiments showed that mixtures of olefins, NO_2, and SO_2 very definitely form particulate matter when radiated by sunlight and that H_2SO_4 was the major constituent of the aerosol particles. It seems very significant that these experiments showed that relatively low concentrations of SO_2; i.e., about 0.10 ppm, were sufficient to promote this H_2SO_4 aerosol formation (30). This means that this reaction can effectively proceed in the ambient atmosphere at the low concentrations which are prevalent in urban areas.

There have been a number of other studies of the photochemistry of SO_2, NO_2, and various hydrocarbons. These include the studies of Harkins and Nicksic (31) who studied reactions between SO_2 and a number of organic materials by using radio tracer techniques. They determined that no organic material was included in the H_2SO_4 aerosols formed by this photochemical system. No experiments with NO_2 were reported, however. Renzetti and Doyle (21) showed that olefins could reduce the rate of photooxidation of SO_2 when NO_2 was not included in the system.

There are no references to indicate that dust or other solid atmospheric aerosols provide any additional mechanisms for SO_2 reaction in the atmosphere. However, it would be logical to expect that there could be situations where gas/solid reaction rates are significant and that SO_2 in the atmosphere could be removed by high concentrations of dust. Alkaline dust would seem to be indicated here and the atmospheric reaction rate would probably be dependent on the rate of reaction between SO_2 and the particle surface in a manner analogous to the reaction rates for alkaline fog drops (van den Heuvel and Mason (25)). Doyle (32) has proposed several mechanisms for SO_2 scavenging by solid particles in aerosol form.

A proven system to explain the reaction of SO_2 in the atmosphere does not exist and is probably beyond the scope of our present knowledge. However, there does not seem to be any doubt that several possible reactions can compete with each other and that the oxidation of SO_2 into H_2SO_4 or a sulfate can proceed relatively rapidly. In summary, it seems that in the daytime and at low humidity photochemical reaction systems involving SO_2, NO_2, and hydrocarbons are of primary importance in the transformation of SO_2 into essentially an H_2SO_4 aerosol. At night and under high humidity or fog conditions or during actual rain it seems that a process involving the absorption of SO_2 by alkaline water droplets and a reaction to form SO_4 within the drop is a well-documented process and can occur at an appreciable rate to remove SO_2 from the atmosphere.

All these studies show how complex the SO_2 oxidation processes within droplets or on aerosols in our atmosphere are, and they also give evidence that the aerosol related processes are of great importance for the removal of SO_2 from the troposphere. Apparently there are several ways in which the oxidation of SO_2 can proceed. However, we are still far from having a clear understanding of which are the most important ones particularly on a global basis.

2. Sulfur Dioxide Reactions with Vegetation and Other Surfaces

The previous discussion dealt only with SO_2 reaction processes occurring totally within the atmosphere.   Reactions of atmospheric SO_2 with vegetation or the ground should also be considered here because they provide an additional mechanism for the removal of SO_2 from the atmosphere.   These reactions provide for the direct removal of SO_2 from the atmosphere and thus are a final sink for the airborne gas.   The atmospheric reactions result in an H_2SO_4 or SO_4 aerosol that must go through an additional deposition process to provide for final removal of sulfur from the atmosphere.

Vegetation absorption of SO_2 and resulting damage from excessive SO_2 concentrations is a well known result of excessive atmospheric concentrations of SO_2.   The rate of SO_2 absorption by vegetation is a complex function of plant growth factors including time of day and weather.   Katz and Ledingham (33) describe experiments using field grown alfalfa covered by fumigation chambers and provided with more than one air change per minute. These experiments emphasize the fact that the SO_2 was rapidly absorbed by the plants.   Air concentrations were in the range of 0.8 to 1.0 ppm SO_2 This same experimental program showed that at night SO_2 absorption by the plants dropped to less than 5 %, while completely shaded plants had maximum absorption rates of around 20 % of the daytime rate.

The result of reactions between SO_2 and plants is an oxidation and fixation of the gas within the plant as sulfate.   Sulfur is thus removed from the atmospheric cycle entirely and therefore it is reasonable to consider this reaction process as a sink for ambient SO_2.

Absorption of SO_2 on the ground and on very low vegetation such as grass also provides a removal mechanism for atmospheric SO_2 which has been studied by Chamberlain (34) as a corollary problem to his interest in the ground deposition of radioiodine.   Chamberlain's method is to calculate a velocity of deposition, v_g, which is defined as the ratio of the rate of deposition per unit surface to the volumetric air concentration.   He then concludes that the material, SO_2 in our case, behaves during deposition processes as though it were a particle with a settling velocity equal to v_g.

In his analysis of SO_2, Chamberlain used two different sets of concentration and deposition data averaged over most of Great Britain, and from these data he calculated v_g values of 0.3 and 0.7 cm/sec.   These values are equivalent to dust ($\rho = 2$ gm/cm³) particles of approximately 7 and 10 μ in diameter.   Since the data he used came from deposit gauges and estimates of the air concentration, no estimates of deposition on plants or absorption be vegetation were used in these calculations.   It is known that dust particles have a finite lifetime in the atmosphere before they return to the earth's surface and if the settling velocity analogy can be accepted, it is one more factor which will indicate a relatively short lifetime of SO_2 in the atmosphere, probably no more than a week or so.

The deposition velocity concept can be applied to the alfalfa experiments described by Katz (33) and mentioned previously. Here the average concentration was about 0.8 ppm or 2,275 $\mu g/m^3$. Since the volume of the chambers was about 8 m^3 (8 × 10^6 cm^3) and the air flow equal to about one change per minute, the amount of SO_2 passing through the chamber was 18,200 $\mu g/min$ or 300 $\mu g/sec$. Maximum deposition was 40%; the deposition area was 4 m^2 (4 × 10^4 cm^2); thus the deposition rate was 3.0 × 10^{-3} $\mu g/cm^2$ sec. The deposition velocity,

$$v_g = \frac{\text{deposition rate}}{\text{air concentration}} = \frac{3.0 \times 10^{-3}\,\mu g\,cm^{-2}\,sec^{-1}}{2.3 \times 10^{-3}\,\mu g\,cm^{-3}}$$

$$= 1.3\ cm\ sec^{-1}$$

The nighttime absorption or "deposition rate" in this expression was about 10% of the maximum, and thus the range of calculated v_g values for Katz's experiments was from about 0.13 to 1.3 cm/sec. In terms of dust particles ($\rho = 2$) this is equivalent to particles between about 5 and 15 μ in diameter. These values are somewhat lower but still compatible with those calculated by Chamberlain (34) for England and will be used when we consider the circulation rate of S from the atmosphere to the land.

It will be pointed out in the next section in which we discuss SO_4 aerosols that the SO_4 particles are primarily in the submicron size. Thus, if Stokes' Law is an adequate guide to the deposition of particulate material, we have the anomolous result that gaseous SO_2 will have a larger apparent deposition rate than will particulate SO_4 where both are present in equivalent concentrations.

Before concluding this discussion of SO_2 reactions with vegetation it should be pointed out that S is a necessary element for growth, and all plants remove S from the soil if it is available (35). In fact, in SO_2 fumigation experiments there may be rare situations where plants improve when exposed to non-damaging levels of SO_2. This action is usually explained by finding that the soil is poor in S. The fact that plants do not usually get large amounts of sulfur from the atmosphere is important when considering the circulation of sulfur through the environment.

3. Reactions of Atmospheric Hydrogen Sulfide

The fate of atmospheric hydrogen sulfide is a problem of considerable importance and interest because of the very large amount of H_2S released into the atmosphere on a global basis. Until a short time ago, little was known of the mode of disappearance of H_2S except that it must be oxidized, ultimately to sulfate aerosol.

In 1961 Robbins (36) examined the reaction of H_2S with ozone and he made the important, but qualitative, discovery that the heterogeneous ozone oxidation on surfaces was very fast. He attempted to measure the homogeneous reaction rate at room temperature, but the rapid surface reaction in the reaction cell dominated the oxidation and the actual rate measured was the heterogeneous rate, although it was not recognized at the time. In 1965 Cadle and Ledford (37) studied the ozone oxidation of H_2S. They confirmed earlier indications that the reaction chemistry was simple,

$$H_2S + O_3 \rightarrow H_2O + SO_2$$

They mixed various ratios of H_2S/O_3 in a flow system and calculated reaction orders from reaction rate measurements at three temperatures. They also varied the surface area/volume ratios and showed that the reaction rate was surface area dependent. From these results they concluded that the reaction was at least partially heterogeneous; also that the reaction was nominally zero order in H_2S and 1.5 order in O_3. Their rate constant was calculated to be:

$$k = 2.5 \times 10^8 e^{-8300/RT} \text{ cm}^{1.5} \text{ moles}^{-0.5} \text{ sec}^{-1}$$

Additional consideration of these reaction studies now leads to an improved model and a rate equation which can be utilized for the calculation of H_2S lifetimes in the atmosphere. With the qualitative confirmation of the very rapid surface O_3 oxidation of H_2S, the surface area/volume ratio measurements of Cadle and Ledford (37) tell us that the reaction rate is a function of surface, or more exactly, of the square root of the specific surface. When they increased the surface area of their reaction chamber by a factor of four, the reaction rate doubled. The rate equation should therefore be written:

$$-\frac{d[H_2S]}{dt} = k[O_3]^{1.5}[A]^{0.5}$$

If concentrations are expressed as moles/cm³, the reaction rate constant, $k = 2 \times 10^2$ cm moles$^{-0.5}$ sec^{-1} at 300° K. [A], the specific area, is actually cm²/cm³; and $[A]^{0.5}$ has the dimensions of cm$^{-0.5}$.

Junge (3) reported that particle concentration in continental atmospheres near the surface is about 15,000/cm³, and the background tropospheric particle concentration is about 200/cm³. The calculated specific areas for these aerosols is 10^{-4} cm²/cm³ and 5×10^{-7} cm²/cm³, respectively. Using these values for [A] and calculating the lifetime of an original one part-per-billion* H_2S exposed to 0.05 ppm ozone, we find for the continental case (15,000 particles/cm³) that the H_2S lifetime is two hours. For the tropospheric background case (200 particles/cm³) the H_2S lifetime is 28 hours.

* 1 part per billion (ppb) = 1 part in 10^9 parts.

Thus, it is obvious that H_2S cannot remain in our atmosphere for very long periods of time, even under the cleanest possible conditions.

4. Atmospheric Reactions of Sulfate Aerosols

The sources of the SO_4 particles found in the atmosphere are the reactions of H_2S and SO_2, which result in an H_2SO_4 aerosol, the evaporation of fog and cloud droplets in which SO_2 was adsorbed and oxidized, and the evaporation of droplets produced by sea spray to leave an aerosol of sea salt which contains a significant SO_4 fraction.

Junge (3) shows that atmospheric aerosols approach a size distribution which can be expressed as a power law of the following form:

$$\frac{dN}{d(\log r)} = cr^{-\beta}$$

in which: N = number of particles smaller than r, r = particle radius, β = slope index, and c = constant. Numerous studies of atmospheric aerosols, including those of Junge (38), have examined the atmospheric aerosol size distribution and in particular the value of the slope index β. Values for β between 2 and 4 have been found to be typical with a value of about 3 being considered typical of an "average aged continental aerosol."

The size distributions of sulfate aerosols, in contrast to that for a general heterogeneous aerosol, have been studied by Junge (38) for uncontaminated atmospheres. Although his data are very limited, they indicate that the tropospheric SO_4 aerosol has 50 % of its mass in particles smaller than about 0.15 μ in radius and that perhaps 25 % of the tropospheric aerosol mass is composed of SO_4 particles. If this is the case, then sea spray can only be a small fraction of the total tropospheric aerosol because the aerosol formation mechanisms for sea spray particles are known to form generally larger particles than found by Junge in these samples.

Since the atmospheric aerosol tends to assume a generally uniform size distribution some mention should be made of the processes which are active in causing this size distribution. These processes have been identified as coagulation, sedimentation or gravitational settling, and the condensation and evaporation cycles followed by cloud, fog, and rain drops (3).

Coagulation of very small aerosols is due to molecular or "Brownian" motion. From a given atmospheric aerosol size distribution Junge and Abel (39) were able to calculate the gradual shift of the size distribution toward larger sizes. Changes were shown to occur most rapidly in the first twelve hours.

Aerosols are also incorporated into cloud, fog, and rain drops. Since this process is a repetitive condensation-evaporation cycle within the cloud or the air mass, numerous small particles can be incorporated into a few larger particles as the cloud droplet sweeps out more and more nuclei. The progress of this process has also been calculated by Junge and Abel (39). The result is a gradual loss of both small and large particles producing a more or less stable size distribution between 0.1 and 1.0 μ.

These aerosol size modification processes will result in a loss of SO_4 compounds from the atmosphere both by gravitational settling and by washout in precipitation. Junge (3) estimates that aerosols in the submicron size range, such as would be characteristic of SO_4 particles in the early stages of formation, will have an atmospheric lifetime of 20 to 30 days and that 80% will be deposited by precipitation and the remainder by "dry fallout." Aerosol deposition will be considered again when the atmospheric sulfur cycle is presented.

The theoretical and laboratory work of Beilke and Georgii (28) has already been mentioned in connection with the washout of SO_2 from polluted atmospheres. These same authors also considered in quantitative fashion the rates of SO_4 aerosol washout by rain. By considering SO_4 aerosols in several size categories, they showed that rain washout efficiencies decreased rapidly with decreasing aerosol particle size, and the probable aerosol washout is almost entirely due to SO_4 particles larger than 0.5 μ in radius. With regard to washout of both SO_4 and SO_2 it should be noted that where there is significant SO_2 in the air mass through which the rain falls, the scavenged SO_2 can be the dominant contributor to the precipitated SO_4.

Calculations were subsequently made by Beilke and Georgii (28) for both the SO_4 rainout resulting from gaseous SO_2 and SO_4 aerosols being incorporated in cloud droplets and the sulfur washout of SO_2 and SO_4 by rain falling through air masses below the cloud mass. This calculation indicates that the sulfate in cloud drops comes mostly from the pickup of SO_4 aerosol particles while the sulfur picked up beneath the active cloud mass by falling rain drops comes from the absorption of SO_2 rather than from the pickup of SO_4 aerosol.

D. Atmospheric Sulfur Concentrations

Sulfur in the form of SO_2 and SO_4 has been measured in polluted atmospheres for many years and voluminous statistics are available. However, in our analysis of the total cycle of sulfur in the environment data are needed on the concentrations of the various sulfur compounds in the clean ambient atmosphere. These data are very sparse, as will become obvious from the following discussion.

1. Sulfur Dioxide

In recent years, some progress has been made in determining SO_2 concentrations in the ambient air, primarily by Georgii and by Lodge and his co-workers. Georgii and Jost (40) succeeded in making the first vertical profiles of SO_2 (and N_2O) over Central Europe and showed that on the average the SO_2 content drops to 1 ppb (i.e., 10^{-9}) or less at altitudes of 5 km. Individual profiles show a great variety of SO_2 decrease with altitude depending on the distribution of inversions and cloud layers. However, these vertical profiles are likely to be affected by the widespread pollution over large parts of Europe and probably show higher concentrations than would occur over unpolluted areas. Recent attempts to obtain such profiles over Nebraska indicate values of less than 0.3 ppb SO_2 in the upper portions of the troposphere (41). These are in line with the few other available measurements of surface SO_2 in very remote places. Older values gave about 0.3 ppb in Hawaii and 1 ppb at the southeast coast of Florida (3). Similar values, of 0.3–1 ppb SO_2, were recently found by Cadle, Pate, and Lodge (42) in Antarctica and in the Panama Canal Zone. Over wide areas of the Central Atlantic, Kühme (43) found no SO_2 above the limit of detection which was about 0.3 ppb. From these figures we have tentatively concluded that the average tropospheric SO_2 concentrations on a global basis is about 0.2 ppb. Further data along the lines of the 1 ppb SO_2 found in the Antarctic by Cadle et al. (44) would no doubt require this estimate to be increased.

This value is considerably lower than previously assumed (e.g., 2 ppb on page 72 of Junge ref. 3), and it is important for estimates of the average lifetime of SO_2 in the troposphere. Using the previous higher average concentration of 2 ppb the following atmospheric residence times for SO_2 were obtained by Junge (3).

$\tau_1 \approx 40$ days if only removal by precipitation is considered, and

$\tau_2 \approx 20$ days if estimates of the direct absorption of SO_2 at the earth's surface are included.

If we now assume a new and probably more realistic concentration of about 0.2 ppb SO_2 the values will be $\tau_1 \approx 4$ days and $\tau_2 \approx 2$ days. It is quite interesting to compare these figures with other data on SO_2 atmospheric lifetimes obtained from regional estimates:

Leicester Report (45)	≈ 4 days
Reevaluation of Meetham's data (3)	≈ 4 days
Estimate based on U.S. data (46) (depending on inclusion of direct soil uptake)	≈ 3–7 days
Estimate by Georgii and Jost (40)	≈ 4 days

With values of τ_1 or τ_2 based on a concentration of 0.2 ppb the older discrepancies largely disappear and all estimates fall in the range of 2–7 days.

Such a scatter could well be real depending on the complexity of the SO_2 oxidation processes, surface absorption, etc., involved in the SO_2 removal. We will subsequently calculate an SO_2 residence time of 3 days from data developed in this report.

Thus, once SO_2 is released into the atmosphere the various reaction processes will act to remove it within a matter of a few days. These processes have already been described and include photochemical oxidation to SO_3, photochemical reactions to form H_2SO_4 in an SO_2–NO_2–olefin system, absorption and oxidation in water drops, and adsorption by vegetation. It seems likely that the more complex photochemical reactions involving SO_2, NO_2, and hydrocarbons are the most generally significant chemical reaction mechanisms for SO_2, while washout by large rain drops seems to be the most effective of the several possible physical mechanisms for SO_2 removal.

2. Hydrogen Sulfide

Although the SO_2 background concentration data are very scarce, there are at least data from several locations. This is not the case for H_2S. Only one set of observations, which were taken in England, has been discovered which might be applicable in estimating background H_2S concentrations. These data are presented by Smith et al. (47) who reported H_2S concentrations of 0.1 and 0.3 ppb in areas of Wales described as "rural areas away from main roads and near the sea shore." Some other data on apparent background H_2S concentrations are summarized by Junge (3), but subsequent unpublished studies by Junge, Lodge, and Stanford Research Institute have shown that the liquid scrubber methods used to obtain these measurements are very doubtful indicators for very low H_2S concentrations. The reason for this seems to be that there is no way to prevent oxidation and loss of significant H_2S. The experiments of Smith et al. (47) also show that the liquid scrubber methods used for H_2S may also suffer from interferences due to SO_2. This would be a significant problem since SO_2 usually seems to be more prevalent than H_2S.

3. Sulfate

Sulfate aerosols complete the spectrum of sulfur compounds in the atmosphere and again the amount of data available for uncontaminated locations is very limited. The data prior to 1966 were summarized by Junge (3) and indicates a background SO_4 concentration of about 5 $\mu g/m^3$. However, the more recent data collected by Junge (38) in remote areas of Oregon indicated SO_4 concentrations in the range of 1.5 to 3 $\mu g/m^3$ with the lowest concentrations being found at higher altitudes and away from the coast. The earlier data were generally coastal samples. Recent SO_4 aerosol data collected by Georgii (41) also indicate that SO_4 concentrations approach 1

$\mu g/m^3$ at altitudes of several kilometers over both Europe and the United States. Since marine air masses are relatively shallow, it appears that an average tropospheric concentration of about 2 $\mu g/m^3$ would be more reasonable for SO_4 aerosols than would a higher value more representative of marine air masses.

4. Stratospheric Sulfate Aerosols

No description of atmospheric S aerosols is complete without some comments on the sulfur aerosol layer in the stratosphere discovered by Junge and his co-workers at the U.S. Air Force Cambridge Research Center (48, 49). As a result of balloon and U-2 aircraft sampling Junge found that aerosol concentrations in the stratosphere increased above the troposphere and exhibited a broad maximum between about 17 and 24 km. Widespread sampling indicated that this was a world-wide phenomenon with very little seasonal variation. In this layer of maximum aerosol the concentration is about four times higher than exists at the troposphere. The size distribution of the particles shows that they range from 0.1 to about 2 μ and that they follow the familiar power law, $\beta = 3.5$, mentioned previously (50).

The chemistry of the particles was studied extensively by Junge (3) and by Friend (50) and it was found that about 80% of the collected aerosol mass consisted of SO_4 and that perhaps 30% of the SO_4 occurred as $(NH_4)_2SO_4$. Traces of Al, Cl, Ca, and Fe also were detected.

The variations in concentration with altitude indicate that the particles were not carried upward from the troposphere by mixing. The rather unique chemistry precludes there being a concentration of extraterrestrial dust. It is Junge's conclusion that this SO_4 layer is formed in the stratosphere, probably from the oxidation of SO_2 or H_2S to form H_2SO_4. After initial formation the particles apparently continue to grow, perhaps with the assistance of some "condensation nuclei," until, when they reach 2 to 4 μ in diameter, gravitation is effective in causing them to fall out of the layer. Junge estimates the residence time for aerosol particles in this stratospheric layer as about 6 months. Martell (51) has put forth another theory of formation for this sulfur aerosol based on the vertical transfer and coagulation of Aitken nuclei. It must be concluded that the particle formation process is still in doubt.

Concentrations in this SO_4 aerosol layer at about 20 km are equivalent to about $4 \times 10^{-2} \mu g/m^3$ (STP) as S. This is about an order of magnitude less than our estimated tropospheric S concentration of either SO_2 or H_2S. The amount of sulfur which cycles through this stratospheric aerosol layer is about 3×10^4 tons per year which is only a very small fraction of the amount circulating through the tropospheric sulfur cycle.

Prior to Junge's sampling studies the presence of a significant stratospheric aerosol layer was postulated by Gruner (52) and others, as an explanation for the purple light which can often be seen in the twilight sky. It was concluded that the purple coloration was due to scattering from the stratospheric aerosols.

5. Concentration Summary

Table VII summarizes our conclusions with regard to atmospheric background concentrations of the various forms of sulfur, and it shows that on the basis of sulfur, about 60 % is in aerosol form, 20 % is present as SO_2, and 20 % as H_2S. This ratio is in general agreement with the conclusions which were reached relative to the rate of reaction of these materials in the atmosphere.

TABLE VII
Average Tropospheric Concentrations of Sulfur Compounds

Material	Average concentration	Average concentration as sulfur, $\mu g/m^3$
SO_2	0.2 ppb	0.25
H_2S	0.2 ppb	0.25
SO_4	2 $\mu g/m^3$	0.7

Hydrogen sulfide is most reactive, and total H_2S emanations are the greatest source of atmospheric sulfur, as indicated in Table VII. Sulfate aerosols are formed primarily from the oxidation of SO_2 and H_2S and since aerosols persist longer than gaseous H_2S or SO_2, the aerosols account for the major fraction of atmospheric sulfur. At present SO_4 aerosol formation from gaseous sulfur compounds is considered much more important than sea spray aerosols as a significant feature of the general tropospheric aerosol. Junge's findings in Oregon (38) support this conclusion.

E. The Environmental Sulfur Cycle

Our present calculations along with additional data from the literature permit us to make an estimate of the circulation of sulfur through our environment. Figure 3 shows our estimate of this sulfur circulation. Some of the values used in this calculation are reasonably well known, i.e., pollutant emissions and total depositions. However some data must be considered very speculative and have been adjusted reasonably to balance the cycle, e.g., land and sea emissions of H_2S.

To better understand this circulation, we will examine its various components. The sulfur annually discharged to the sea by the world's rivers is

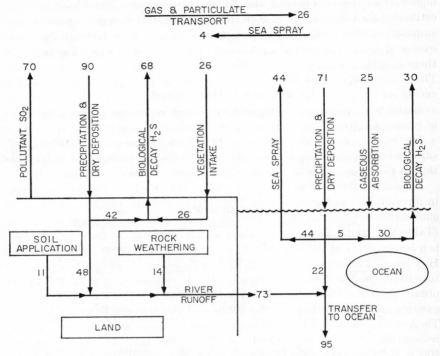

Fig. 3. Environmental sulfur circulation (units: 10^6 tons/yr sulfur).

73×10^6 tons; this results from sulfur accumulated from weathering rocks, 14×10^6 tons; sulfur applied to the soil in fertilizer, etc., 11×10^6 tons; and sulfur deposited on the soil by precipitation or dry deposition, 48×10^6 tons. These amounts were determined by Eriksson (2).

The atmosphere-land portion of the cycle contains 70×10^6 tons of pollutant sulfur emitted to the atmosphere, 90×10^6 tons of sulfur deposited from the atmosphere to the land, a loss of 68×10^6 tons of sulfur as H_2S from decaying vegetation, and an intake of sulfur by vegetation from the atmosphere of 26×10^6 tons. The 90×10^6 tons deposited includes 80%, or 70×10^6 tons, in rain and the remainder as dry deposition. These follow from estimates presented by Junge (3). As indicated, 48×10^6 tons of this deposited sulfur is carried off by rivers, and 42×10^6 tons is absorbed by vegetation and then released as H_2S. The intake of sulfur by vegetation is estimated to be 26×10^6 tons based on calculations using a deposition velocity of 1 cm/sec and an ambient concentration of 0.4 ppb (0.5 $\mu g/m^3$). This is twice the average tropospheric concentration listed in Table VII using the argument that ground level concentrations over land would be higher than the average for the whole troposphere. The 68×10^6 tons estimated for the emission of S as H_2S from vegetation decay results from a summation of the atmospheric vegetation intake, 26×10^6 tons, and the excess of deposition over river carryoff, 42×10^6 tons. This assumes that there is no net accumulation in the surface soils, which seems reasonable. This value is close to Eriksson's (2) estimate of 77×10^6 tons for H_2S emissions from land areas; however, this is a value for which no data are available with which it can be checked. Land areas also gain 4×10^6 tons of S from sea spray (2).

The net result of this land circulation is an excess of 26×10^6 tons of sulfur which must be deposited in the ocean if there is to be no net accumulation in the atmosphere.

Deposition of sulfate in the ocean is 71×10^6 tons (3) with 80% contained in precipitation and 20% as dust. The ocean also absorbs 25×10^6 tons of gaseous S calculated on the basis of an average SO_2 concentration of 0.2 ppb (Table VII) and a deposition velocity of 0.9 cm/sec (2). The ocean surface is a source of 44×10^6 tons of S in sea spray (2) and 30×10^6 tons of S as H_2S from vegetation decay. There is a tropospheric transfer of 4×10^6 tons of S from the ocean to the land. The H_2S emission of 30×10^6 tons is obtained on the basis of what is needed to balance the 100×10^6 tons of gaseous and solid pickup by the ocean and the transfer from sea to land. There are no data which would provide a check as to whether or not this is reasonable. It is a significantly smaller value than the approximately 200×10^6 tons estimated by Eriksson (2) and Junge (49) for similar calculations.

The end result of this cycle is an accumulation of S in the oceans of 95×10^6 tons which is the sum of pollutant emissions, S applied to the soil, and rock weathering.

In many ways this cycle is similar to sulfur cycles derived by Eriksson (2, 53) and Junge (49). However, there is one major and significant improvement. This is the provision for a net transport from the land to the ocean which is in agreement with all the observed concentration patterns. The cycles derived by both Eriksson and Junge had a net transfer of sulfur from ocean to land in contradiction to the data, as was recognized by Junge (49). The reason our cycle is different is in the considerations of the biological cycle where both Junge and Eriksson equated gaseous plant intake from the atmosphere, 26×10^6 tons in our case, to the H_2S release from decaying vegetation. It is recognized that plants can absorb considerable S from the soil and that gaseous S intake is not a necessary factor to account for S in plant materials. Thus, in our cycle we used the deposition velocity concept to determine plant intake and a balance between deposition, river runoff, intake, and H_2S emission to establish the latter term. This

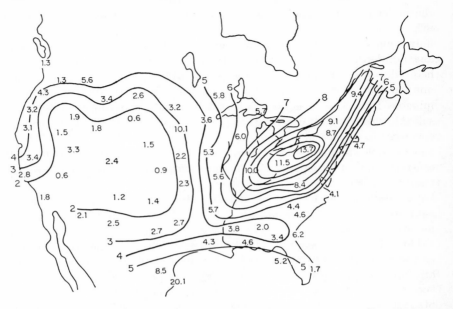

Fig. 4. Excess sulfur in precipitation over the United States (units: $g/m^2/yr$) (Eriksson, 1960).

procedure permitted the land area to be a significant source of S and to provide the desired land to ocean transport.

Data on S deposition over the United States and northern Europe have been analyzed by Eriksson (2), Junge (46), and by de Bary and Junge (54). In Figure 4 [Eriksson (2)] the annual deposition of nonmarine S over the United States shows a very high area in the Midwest and Northeast. This pattern clearly follows the prevailing wind pattern and must obviously produce significant S deposition for a considerable distance over the Atlantic Ocean even though there is apparently a rapid decrease at the east coast in this average pattern. It is interesting to note that west of the Mississippi River in the United States the deposition of S is relatively low but still present in measurable quantities. This S deposition is probably due more to natural emissions of H_2S than to SO_2 in view of the fact that SO_2 emissions are neither large not widespread in this area where natural gas rather than coal or oil is the dominant fuel.

Deposition over the ocean is reasonably effective because the nonmarine SO_4 deposition in the Atlantic coastal area of Europe is relatively low, and in Europe SO_4 concentrations increase generally as the air masses penetrate farther inland (54). Over Europe the S pattern clearly shows the effects of the major industrial areas and their large-scale anthropogenic emissions of SO_2.

It would be very desirable if the S cycle could indicate whether there were any net accumulation of S in the atmosphere as well as that indicated as being deposited in the ocean. Unfortunately, the data are neither precise nor complete enough to permit any such evaluation. In an effort to determine whether there has been a relatively recent increase in atmospheric sulfur, Junge (46) analyzed ice samples from the Greenland Ice Cap which covered a dated period from 1915 to 1957. In this relatively short span he was not able to detect an increase in deposited sulfur. He concluded that the relatively rapid loss of S from the atmosphere in the several days in transit to northern Greenland prevented the detection of any indications of an increase in sulfur due to pollution emissions. If ice samples could be obtained over a longer period, i.e., 100 to 200 years, some secular changes might be detected.

This environmental sulfur circulation can be evaluated by comparing its features with other circulation factors which we have estimated, namely, the atmospheric residence times for SO_4, SO_2, and H_2S. These have already been mentioned in our discussion as being approximately 20 to 30 days for SO_4, four days for SO_2, and "considerably shorter," perhaps one day, for H_2S.

In the annual sulfur circulation it is recognized that atmospheric SO_4 comes from several sources: sea spray, 44×10^6 tons S; oxidation of SO_2, 70×10^6 tons less the 51×10^6 absorbed by vegetation and the sea surface for a net 19×10^6 tons S; and oxidation of H_2S, 98×10^6 tons. This results in a total of 161×10^6 tons S in the form of SO_4 which annually circulates through the atmosphere. Now, in Table VII we determined that a reasonable average concentration of SO_4 was 2 $\mu g/m^3$ or 0.7 $\mu g/m^3$ S which may also be expressed as 0.6 ppb S, by weight. Considering the total mass of the earth's atmosphere as 5.3×10^{21} g or 5.7×10^{15} tons means that the total average amount of S in the atmosphere as SO_4 at any one time is 3.5×10^5 tons. Comparing this amount to the 161×10^6 tons of S annually circulating through the atmosphere as SO_4 indicates that the atmospheric SO_4 is replaced 46 times in a year for an average residence time of about eight days. Although this is shorter than the estimated SO_4 cycle based on estimates of the cycle of dust particles, it is within the range of values calculated for sulfur in urban source regions. Considering the various approximations in this estimation, this value of eight days for the SO_4 cycle must be considered as reasonable and consistent with the known features of the atmospheric system.

Similar calculations can be made for both SO_2 and H_2S. In the case of SO_2 the annual production is 70×10^6 tons as sulfur. Given an average atmospheric concentration of 0.2 ppb SO_2 or 0.1 ppb, by volume, as sulfur in an atmosphere of 5.7×10^{15} tons, the amount of sulfur as SO_2 in the atmosphere at any given time is 1.2×10^6 tons. Therefore, the 70×10^6 tons produced annually must pass through about 60 cycles in a year, which is an indicated atmospheric residence time of six days for SO_2. This is close to our previously stated average of 4 days.

Residence time calculations for H_2S are less sure than for either SO_4 or SO_2 because of our general lack of information on concentrations and general production. However, our circulation calculations indicate a production of 98×10^6 tons annually. Based on the estimated average H_2S concentration of 0.2 ppb reported in Table VII, about 10^6 tons of sulfur is present as H_2S. Thus, the estimated annual production is about 100 times the atmospheric amount, and the indicated H_2S residence time is less than four days. This is somewhat longer than was indicated by the chemical reactivity of H_2S in the atmosphere, but it is probably within a factor of 2 or 3 of the correct value.

These SO_4, SO_2, and H_2S residence times are all reasonably close to values arrived at by considering other parameters of the system. It should also be pointed out that SO_4 particulate material, which would be expected to be least reactive, had the longest residence time. These calculations are interpreted as an indication that we are probably not seriously in error in our estimates of the environmental sulfur circulation.

It is tempting to try and draw further conclusions from these calculations, i.e., to estimate whether H_2S emissions provide a source, through an atmospheric reaction, for a significant steady state concentration of SO_2, or does the H_2S reaction go rapidly to SO_4. This and other speculations seem to be pushing our estimates and limited data much too far.

F. Discussion of Sulfur Compound Studies

In most evaluations of long-term and large-scale atmospheric systems it is important to see whether any measurable long-term changes have occurred. Although Junge's attempt to measure SO_4 deposition changes in Greenland over a 40-year span failed to show any change, there are some other measurements of particulate materials which appear to indicate a secular increase in general atmospheric particulate concentrations.

An increase in atmospheric fine particle concentrations over the period from 1910 to 1962 was postulated by Gunn (55) on the basis of electrical conductivity measurements over the oceans. In the period from 1910 to 1929 conductivity measurements were taken aboard the research ship "Carnegie." The values determined in 1928 and 1929 have been corrected by Gunn for the effect of turbulent eddy diffusion in the detection instrument used on these two occasions. In prior years the instrumentation, while not entirely satisfactory by modern standards, did not have the serious errors induced by unsuspected turbulence.

Conductivity measurements of this type are most closely related to fine particles in the Aitkin nuclei range (particle radius from 2×10^{-7} cm to 10^{-4} cm). Sulfur is a siginficant component of these particles. Particles in this range act as a sink to remove from the atmosphere the small ions which are responsible for the electrical state of the atmosphere. The greater the concentration at Aitkin nuclei, the lower the conductivity because the concentration of small ions is reduced. It is not possible to give quantitative values for the increase in the mid-ocean fine particle concentration, but there does not seem to be much doubt, because of the reduction in conductivity, that it has taken place. Furthermore, there seems to be little doubt that this increase in the fine particle concentration is a direct result of the increased pollution load which has been forced into the atmosphere.

The data available can be separated into Atlantic and Pacific measurements. Except for the two earliest years, 1928 and 1929, the Atlantic measurements all show lower conductivity than do the Pacific measurements. It can be inferred from this that the Atlantic air masses carry a greater load of fine particles than do Pacific air masses. This is not an unreasonable conclusion considering the fact that the Atlantic is both smaller than the Pacific and generally dominated by heavily industrialized North America. The observed change in the fine particle concentration seems to be small and it is probably not yet a significant factor in atmospheric processes. However, the fact that the changes have been great enough to be detected at all should be of concern to serious investigators in the air pollution field. It should also be mentioned here that McCormick and Ludwig (56) have postulated that an increase in atmospheric turbidity has occurred in the past 40 to 50 years, and that this could be attributable to an increase in fine particle concentration. They also point out that a significant and persistent increase in atmospheric turbidity, even a relatively small change, could cause a change in long-term climatic cycles.

Sulfate is a major component of the atmospheric aerosol and there have been steady increases in S emissions over the past years. Thus, it would be difficult to eliminate SO_4 aerosols from being considered as one of the causative factors in these long-term atmospheric changes attributed to increases in the atmospheric fine particles.

This compilation of facts and discussions about sulfur in our environment points out a number of interesting ideas especially relative to pollutant and natural sources of sulfur. With regard to estimates of gaseous sulfur emissions, our evaluation of available data indicates that natural emissions of S, in the form of H_2S, are about 30% greater than are the estimated industrial emissions of SO_2 and H_2S, i.e., 100×10^6 tons compared to 76×10^6 tons. With regard to sulfur pollutants, the most important fact is that SO_2 is the only significant pollutant and the transformation of SO_2 to SO_4 in the form of H_2SO_4 occurs quite rapidly. As a result of this rapid reaction the expected atmospheric life of a volume of SO_2 is only a few days, perhaps about four. Most of the emitted SO_2 becomes SO_4 in the atmosphere as a result of several possible photochemical or physical reactions. This rapid reaction rate plus ready absorption of SO_2 by vegetation contributes to a rapid decrease in concentration outside emission source areas.

In the ambient troposphere the majority of the sulfur is present as SO_4. This results from the rapid oxidation of SO_2 and H_2S and from the large amount of SO_4 aerosol resulting from sea spray generation on the ocean surface.

Unfortunately, the data on atmospheric sulfur concentrations, whether SO_2, H_2S, or SO_4, for areas beyond the direct influence of pollution sources are very sparse. It is unlikely that we can adequately evaluate the circulation of sulfur and the relative importance of the various sulfur compound sources until a considerable amount of additional data is gathered over the oceans and remote land areas of the world.

III. CARBON DIOXIDE IN THE ATMOSPHERE

A. Introduction

Air pollution analyses are typically concerned with compounds that are only present in the atmosphere in low concentrations and that pose well-recognized adverse reactions. This generality does not hold for the most commonly emitted pollutant, carbon dioxide. Carbon dioxide (CO_2) is an integral factor in the life cycle of the earth itself. At present, even though pollutant emissions of CO_2 are very great, the amount of CO_2 that is cycled through the biosphere is much greater, and there are several logical explanations as to why CO_2 is often neglected when pollutant emissions are tabulated.

As far as air pollution control is concerned, CO_2 is so common and such an integral part of all of our activities that air pollution regulations typically state that CO_2 emissions are not to be considered as pollutants. This is perhaps fortunate for our present mode of living, centered as it is around carbon combustion. However, this seeming necessity, the CO_2 emission, is the only air pollutant, as we shall see, that has been shown to be of global importance as a factor that could change man's environment on the basis of a long period of scientific investigation. Because of this obvious relation, we believe that any discussion of atmospheric pollutants should also include a discussion of CO_2.

The possibility that changes in atmospheric CO_2 could change world climate is not a new idea.   It was first proposed independently in America by Chamberlain in 1899 and in Sweden by S. Arrhenius in 1903.   Since then, CO_2 and possible geophysical effects have been the source of much discussion and investigation.   At present, while we know a great deal more about the problem, we still cannot quantitatively evaluate the impact of the accumulation of CO_2 in the atmosphere in terms of climatic change.   Considerable research is under way in a number of laboratories on this problem, however.

As a basis for our discussion, the essential features of the CO_2 pollution problem can be briefly outlined.

1. CO_2 emissions are extremely large and there has been for many years a gradual increase in atmospheric CO_2 concentrations.

2. Although there are sinks for CO_2 in both the marine and biospheric environments, none of them is capable of counterbalancing the increasing pollutant emissions.

3. The impact of CO_2 on the environment is through its interaction with the earth's radiation balance.   This is a very complex relationship, and the magnitude of possible environmental changes due to CO_2 are not now clearly defined.

B. Atmospheric Carbon Dioxide Emissions

Carbon dioxide is present in the uncontaminated atmosphere at an average concentration of about 320 ppm.   The source is the oxidation of carbonaceous materials in processes of respiration by both plants and animals, decay of organic materials, and the combustion of organic fuels.   If there were no sinks for CO_2 it would obviously continue to increase; however, at least until recently, there was a balance between the release and the consumption of CO_2 (by photosynthesis in plants, and by the formation of calcium carbonate). This balance between environmental sources and sinks has been disturbed by the emission to the atmosphere of additional CO_2 from the increased combustion of carbonaceous fuels.

The amount of CO_2 emitted from pollution sources can be estimated on the basis of the carbon content of the fuel.   Table VIII is an estimate of CO_2 emissions based on estimates of 1965 fuel usage.   The total is 14.08×10^9 tons (12.8×10^{15} g).   More than 50% of the total CO_2 results from coal and other solid fuels; petroleum combustion produces about 30% of the total; and natural gas less than 10%.   The remainder is due to miscellaneous forms of combustion such as forest fires, incineration of wastes, etc.

On the basis of the fact that the major commercial fossil fuels (coal, petroleum, and gas) account for 89% of the estimated total, it is possible to estimate the change in emission rates over the past 100 years.   Figure 5 shows such an estimate based on decade averages for fuel consumption for the

TABLE VIII

Estimated Global CO_2 Emissions Based on 1965 Fuel Usage

Fuel	Usage	CO_2 Factor	CO_2 Emission (tons)
Coal	3074×10^6 T	4770 lb/T	7.33×10^9
Petroleum			
Gasoline	379×10^6 T		
Kerosene	100×10^6		
Fuel oil	287×10^6		
Residual oil	507×10^6		
	1273×10^6 T	6320 lb/T	4.03×10^9
Natural gas	20.56×10^{12} ft^3	0.116 lb/ft^3	1.19×10^9
Waste incineration	500×10^6 T	1830 lb/T	0.46×10^9
Wood fuel	466×10^6 T	2930 lb/T	0.68×10^9
Forest fires	324×10^6 T	2380 lb/T	0.39×10^9
Total			14.08×10^9 T
			$(12.8 \times 10^{15}$ g)

period 1860–1960 (57).   These data are lower estimates of the CO_2 emissions because they do not take into consideration considerable amounts of so-called noncommercial fuels such as wood, wood waste, bagasse, etc.   In 1952 these fuels accounted for an estimated emission of 1.4×10^{15} g of CO_2 (58). In earlier years the per capita consumption of these fuels was undoubtedly greater, but estimates of total usage are not available.   However, even if the nineteenth century data are low because of unaccounted for fuel, there is

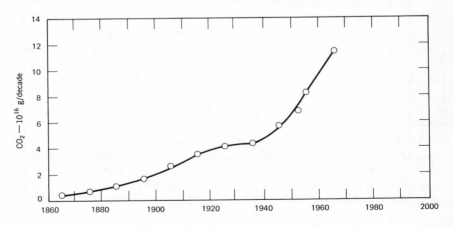

Fig. 5. Average CO_2 emissions to the atmosphere from fossil fuel combustion.   Decade average data 1860–1960.

still no doubt that there has been a significant increase in the atmospheric emissions in the past 100 years.   The increase has probably been between 6 and 10 times.

Over the period covered by Figure 1 the total production of CO_2 from fossil fuel combustion has amounted to 32.44×10^{16} g (57).   This is equal to 13.5% of the toal CO_2 in the atmosphere, which was estimated from 1957 to 1959 data by Takahashi (59) to be 2.41×10^{18} g.

C. Atmospheric Carbon Dioxide Concentrations

As previously mentioned, the present estimate of the average CO_2 concentration in the atmosphere is about 320 ppm.   Fluctuations about this average are quite large, depending on the nature of nearby sources and sinks. For example, during the growing season, concentrations decrease because plants us up CO_2 during photosynthesis.

One of the initial investigators of the possibility of CO_2 changes in the atmosphere in the 1930's was Callendar, and in 1958 he published an analysis of historical CO_2 measurements covering the period from 1870 to 1955 (60). His results are shown in Figure 6 and indicate a nineteenth century CO_2 base concentration of about 290 ppm.   This value is generally accepted as a reasonable approximation of the CO_2 concentration in the undisturbed atmosphere.

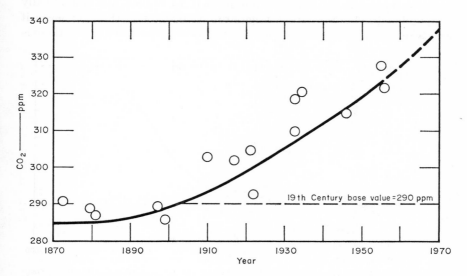

Fig. 6. Average CO_2 concentration in North Atlantic Region 1870–1956.   (Callendar, G. S., *Tellus* **10**, 243, 1958).

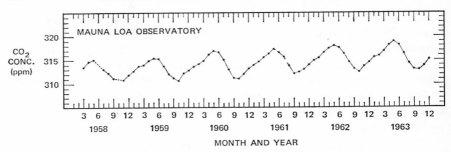

Fig. 7. Monthly average concentration of atmospheric CO_2 at Mauna Loa Observatory versus time.   (Pales, J. C., and C. D. Keeling, *J. Geophys. Res.* **70,** 6066, 1965).

Figures 7 and 8 show CO_2 concentration data from Mauna Loa, Hawaii for the 6-year period 1958–1963 (61).   These data have been screened to eliminate local biasing effects, and the results should represent average conditions in the Northern hemisphere over this time period.   Figure 7 clearly shows the annual cycle of CO_2 concentrations, with maximum concentrations occurring in the late spring and minimums in the fall.   A gradual trend toward rising concentrations is identifiable in Figure 7, but Figure 8, which is a plot of the 12-month running mean concentration, shows the upward trend much more clearly.   As shown by Figure 8, the CO_2 concentration has increased from about 313.3 ppm for the 12 months before June 1959 to a concentration of 315.9 ppm for the 12 months before June 1963.   The straight line fitted to the data of Figure 8 indicates an annual rate of increase of 0.68 ppm.

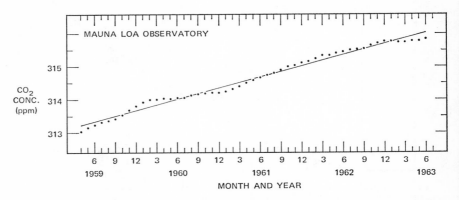

Fig. 8. Twelve-month running mean of the concentration of atmospheric CO_2 at Mauna Loa Observatory.   Means are plotted versus the sixth months of the appropriate 12 month interval.   The straight line indicates a rate of increase of 0.68 ppm/yr. (Pales, J. C., and C. D. Keeling, *J. Geophys. Res.* **70,** 6066, 1965).

There is an apparent difference of several ppm between these Hawaiian data and the 1955 results reported by Callendar in Figure 6. This difference is probably due to various local effects resulting from land-based sampling in the case of Callendar's data. The Hawaiian data and other similar observations from locations where good exposure to the free atmosphere is attainable probably best represents the present state of the well-mixed troposphere.

The rate of increase shown by the Mauna Loa data seems to be charac-teristics of global CO_2 concentrations. Bischof and Bolin (62) show that Scandinavian data fit the postulated 0.7-ppm/year rate obtained from the Mauna Loa data. Figure 9 shows average monthly data from the Antarctic for the period from September, 1957 to November, 1959 (63). The rate of increase here is about 1 ppm/year, which, considering the limited span of the data, seems to be comparable to the Northern hemisphere data. The average 1959 Little America concentration of about 314 ppm is also com-parable to the Mauna Loa average of 313.3 ppm indicated by Figure 4.

Around 1960, CO_2 concentrations were increasing by 0.7 ppm or 0.22% of a 315 ppm average concentration, and in this same period emissions from fossil fuel, as shown in Figure 5, were adding about 10^{16} g of CO_2 per year to the atmosphere for a 0.42% rate of addition. Thus, it appears that very nearly half the CO_2 emitted to the atmosphere remains there. Revelle (57) came to a similar conclusion.

The data on atmospheric CO_2 are quite voluminous and it is not possible to present them in detail in this discussion. However, the data indicate

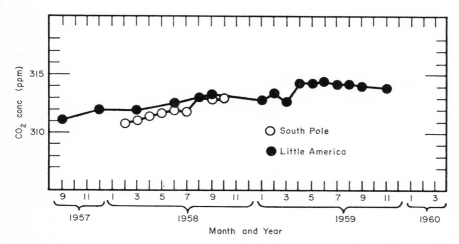

Fig. 9. Variations in CO_2 concentrations in the antarctic (C. D. Keeling, *Tellus* **12,** 200, 1960).

that any differences between the hemispheres are small and that there has been a readily detectable increase of CO_2 in the atmosphere. This increase was originally postulated on the basis of an analysis of nineteenth century measurements. The increase from a nineteenth century base of 290 ppm to a late 1960 average of about 320 ppm is 10%.

D. Scavenging Mechanisms for Atmospheric Carbon Dioxide

The previous discussion has mentioned that processes in the biosphere and the oceans act to maintain an equilibrium situation between the atmosphere, the biosphere, and the oceans.

In the biosphere, photosynthesis removes CO_2 from the atmosphere and forms plant carbon. The ultimate fate of most of this carbon is deposition as litter and humus followed by decay and the rerelease of CO_2 to the atmosphere. Thus, the cycle is completed.

Detailed studies of the relationship between atmospheric CO_2 and the biosphere have been made. Lieth (64) pointed out that the exchange between the atmosphere and the biosphere is very intensive through the two processes of assimilation and respiration. The total amount of carbon uptake by land plants has been estimated for 10° latitude belts by Junge and Czeplak (65) using data developed by Lieth (66). These data, in terms of a global distribution of CO_2 uptake, are given in Table IX. The total annual rate of CO_2 vegetation uptake is 141×10^{15} g, with 83×10^{15} g or 59% occurring in the Northern hemisphere. If vegetation uptake was the only scavenging mechanism for the 2.41×10^{18} g of CO_2 in the atmosphere, the residence or turnover time for CO_2 would be 17 years. This estimate is based on the estimate of the total vegetation production capacity of the world shown in Table IX. In many references the CO_2 residence time resulting from vegetation scavenging is often indicated as being about 30 years and this represents understandable variations in estimates of the total production of the biosphere.

It is obvious that there is considerable variation in CO_2 production and uptake rates within various vegetation communities. For example, Lieth (64) estimates that a tropical forest in Malaya has an annual productivity

TABLE IX

Total CO_2 Uptake by Land Plants*

(10^{15} g year)

Latitude (°)	0–10	10–20	20–30	30–40	40–50	50–60	60–70	70–90
Northern hemisphere	23.5	14.3	11.7	9.2	10.3	9.2	4.8	0
Southern hemisphere	26.4	18.0	8.8	4.0	0.7	0	0	0

* Junge and Czeplak (65); Lieth (66).

equivalent to 37.5 metric tons per 10^4 m^2 of CO_2, and that this is equivalent to a turnover time for the CO_2 above this tropical forest of 0.8 years. A similar estimate for a forest in Europe resulted in an annual productivity rate equivalent to 15 metric tons per 10^4 m^2 of CO_2 and a CO_2 turnover time of 2 years.

Although the biosphere is important in the CO_2 cycle, the role of the ocean in the CO_2 cycle has long been recognized as the dominant one, and various estimates of the average atmospheric residence time for CO_2 relative to the uptake rate of the oceans have been calculated. Some of the more detailed calculations have been made by Young and Fairhall (67) using data on bomb-produced C^{14}. If only the troposphere is considered, their estimate is that the CO_2 uptake rate of the ocean is sufficient to give a mean residence time of CO_2 in the atmosphere of 2.5 years. This average value is made up of two residence times, 3.9 years for the Northern hemisphere and 1.9 years for the Southern hemisphere.

This mean tropospheric residence time of 2.5 years and the tropospheric CO_2 mass of 202×10^{16} g, 84% of the atmospheric total, indicates a global oceanic uptake rate of 81×10^{16} g/year. If this is the only sink for the CO_2 in the stratosphere as well as the troposphere, an average atmospheric CO_2 residence time of 3 years is indicated. Other calculations for the CO_2 atmospheric residence time have varied from less than 2 years to about 20 years (67).

Within the ocean, CO_2 is consumed in the growth of marine plants and animals just as it is in the biosphere. The production of CO_2 during the decay of this material is a major source of both atmospheric and marine CO_2. The CO_2 that is used by marine organisms is primarily used to build the organic mass of the organism, and partly to generate shells and other calcareous material.

Biological activity including CO_2 takes place for the most part in the upper 50 to 100 m of the ocean. This is the mixed layer. However, obviously as these organisms die there is considerable settling of material out of the mixed layer and into the deep water layer. This settling action is a major sink for carbon and atmospheric CO_2 because the deep water is essentially isolated from the atmosphere. The mixing time for the deep water has been estimated to be upwards of 500 years.

In the late 1950's there was considerable new scientific discussion about the secular increase of CO_2, as a result of published studies of atmospheric levels by a number of researchers, including Revelle and Suess (68) and Callendar (60), and predictions of resulting temperature changes in the atmosphere by Kaplan (69), Plass (70), and others. Initially in these discussions the correctness of Callendar's (60,71) estimate of about a 10% secular rise since 1900 in atmospheric CO_2 was discounted (see, e.g., 68,72) primarily on the assumption that there would be relatively rapid exchange of CO_2 between the oceans and the atmosphere. Since CO_2 is highly soluble, one argument was that because the oceans are a reservoir of CO_2 that is very large compared with the atmosphere (about 60 times as much CO_2 is stored in the oceans as in the atmosphere) equilibrium between the ocean and the atmosphere would be achieved rapidly enough to minimize any significant atmospheric CO_2 accumulation due to combustion emissions. Initial data on $^{12}C/^{14}C$ relationships in the atmosphere and the oceans, the so-called "Suess effect," also indicated that combustion-derived CO_2 was not very significant in any change in atmospheric CO_2 levels (68).

More information became available in the late 1950's, however, and the subsequent detailed analysis of the atmosphere-ocean system showed several important things relative to CO_2 buildup. First, IGY sampling data showed that Callendar's postulated increase of CO_2 was occurring and at a rate of about 0.7 ppm/year, as we have mentioned. Second, careful analysis of the ocean/atmosphere exchange processes indicated that only the relatively shallow mixed layer was important and that this layer provided very little in the way of storage to moderate any secular increase of atmospheric CO_2.

Bolin and Eriksson (73) show that this low storage factor is due to a buffering mechanism set up in the sea by dissolved CO_2. It is related to the dissociation equilibrium between CO_2 and H_2CO_3 on the one hand and HCO_3 and CO_3 ions on the other. The net result is that a 10% change in atmospheric CO_2, such as from 290 to 320 ppm, would be balanced by only a 1% increase in the CO_2 content of sea water. This buffering mechanism was originally pointed out by Revelle and Suess (68), but it was unimportant in their analysis because in their ocean model it was applied to the whole volume of the ocean. Bolin and Eriksson (73), using the more realistic two-layer model of the ocean, showed that when this buffering mechanism, acting in the mixed surface layer, was taken in account, there was good agreement between the observed and postulated CO_2 changes in the atmosphere due to fossil fuel combustion. They also showed that when changes in dissociation equilibrium resulting from CO_2 transfer were accounted for, $^{12}C/^{14}C$ ratios in the atmosphere and the oceans were in agreement with an increase in atmospheric CO_2 of about 10%.

Various authors have from time to time attempted to relate atmospheric CO_2 changes to various changes in the physical properties of the ocean or to changes in the biosphere. Eriksson (53) made a comprehensive analysis of the possible changes in CO_2 that might result from changes in oceanic conditions. His conclusion was that if the temperature of the oceans increases by $1°$ C, the atmospheric CO_2 increases by about 6%. If only the mixed layer of the ocean is considered, data given by Junge and Czeplak (65) permit calculation of the effect of a limited temperature change of surface water. The result is that if the mixed layer alone increased in temperature by 1° C, the atmospheric CO_2 would increase by about 0.4%. We conclude with regard to recent increases in atmospheric CO_2 that any reasonable changes in oceanic temperatures could only result in local fluctuations in atmospheric CO_2 and not in any changes of global importance.

Other changes such as water volume, seawater chemistry, and variations in the marine biomass may also be discounted as contributors to short term changes in atmospheric CO_2 levels; although these factors may be significant in climatic changes over a geologic time scale (53).

Changes in the biosphere are also sometimes linked to changes in atmospheric CO_2. It is pointed out that vegetation responds with increased production under increased concentrations of CO_2 and that an increase in the mass of the biosphere could be a moderating influence on any increase in atmospheric CO_2.

A rapid increase in the production rate of the biosphere could also cause an increase in atmospheric CO_2, at least until a new stable position was established. Data on the mass of material in the biosphere is relatively crude; however, it seems unlikely that the observed rise in atmospheric CO_2 has been due to changes in the biosphere.

E. Effects of Atmospheric Carbon Dioxide Concentrations

Our discussion has already indicated that atmospheric CO_2 concentrations are of concern because of the effect of atmospheric CO_2 on the radiation balance of the earth. The effect of CO_2 is a result of the interception of long-wave radiation from the earth's surface and its reradiation to the surface. Under the simplest set of assumptions for this situation the earth's surface reaches equilibrium at a higher temperature, and this is recorded as in increase in air temperature.

Under more realistic conditions, it is recognized that this situation is not a simple relationship. One of the major factors is an interaction between the CO_2 and water vapor in the atmosphere. Water vapor is significant because it also absorbs strongly in the infrared. Cloudiness, because it intercepts incoming radiation, is also an important parameter. However, the most important interaction is between the radiation regime and the general atmospheric circulation, since it is the circulation pattern that determines the weather.

One of the most detailed analyses of the temperature effect from CO_2 changes has been carried out by Manabe and Wetherald (74). This model is based on an atmosphere that is in radiative convective equilibrium, and the model approaches the real situation by estimating temperature effects for the atmosphere itself instead of for the earth's surface as had prior models.

In addition, previous models such as that used by Möller (75) had usually been especially sensitive to some specific sets of conditions and produced very different end results with relatively minor changes in some critical assumptions. Manabe and Wetherald claim that this does not occur in their model, and thus they argue that they have probably come closer than previous writers to approximating the real situation.

A number of different sets of conditions were used by Manabe and Wetherald (74), but probably their most representative model is one in which relative humidity is maintained constant and an average amount of cloudiness is included. The result is that for a change in CO_2 from 300 to 600 ppm, an increase in temperature of 2.36° C is estimated. For a CO_2 decrease from 300 to 150 ppm, a temperature decrease of 2.28° C is predicted. These data indicate that an increase to 330 ppm would produce an estimated temperature increase of about 0.3° C. In summarizing their CO_2 modeling experiments, Manabe and Wetherald point out the following specific factors:

1. The higher the CO_2 concentration, the warmer the equilibrium temperature of the earth's surface and the troposphere.

2. The higher the CO_2 concentrations, the colder the equilibrium temperature of the stratosphere.

3. Relatively speaking, the stratosphere responds much more strongly to a change in CO_2 content than does the troposphere.

The importance of these CO_2 model calculations is that we expect that the CO_2 concentration in the atmosphere will continue to increase as our combustion economy continues to consume increasing amounts of fossil fuel. Various estimates have been made about the possible CO_2 that might be released to the atmosphere. Table X is an estimate of CO_2 emissions from the present to the year 2000. The base point for this information is our estimate for 1965 in Table VIII and estimates of global changes in fuel use. The estimated increases for coal (0.2%/year), petroleum 6.2%/year), and natural gas (7.2%/year) are based on recent estimates of growth to 1980, extrapolated to 2000 (19). For the minor sources, the change in incineration is intermediate between the major fuels, and no changes are estimated for the burning of fuel wood or for forest fires. On a relative basis, CO_2 emissions in the year 2000 are almost 3 times higher than those in 1965. The estimated total CO_2 emission over the period 1965–2000 is 845×10^9 tons or 770×10^{15} g. If half the emitted amount remains in the atmosphere, as we previously calculated, the total CO_2 in the atmosphere would be about 2.8×10^{18} g or about a 16% increase from 320 ppm in 1960 to about 370 ppm in 2000. On the basis of Manabe and Wetherald's (74) calculations, a 50 ppm increase would be translated into an average atmospheric temperature increase of 0.5° C.

TABLE X
Projected CO_2 Emissions: 1965–2000

	1966 Usage	CO_2 Factor	Emissions, 10^9 T				
			1965	1970	1980	1990	2000
Coal	$3,074 \times 10^6$ T	4770 lb/T	7.33	7.40	7.55	7.70	7.85
Increase	0.2% yr						
Petroleum	$1,273 \times 10^6$ T	6320 lb/T	4.03	5.28	8.57	13.90	22.50
Increase	6.2% yr						
Natural gas	20.56×10^{12} ft^3	0116 lb/ft^3	1.19	1.62	2.79	4.80	8.27
Increase	7.2% yr						
Incineration	500×10^6 T	1830 lb/T	0.46	0.51	0.61	0.73	0.88
Increase	2% yr						
Wood fuel	466×10^6 T	2930 lb/T	0.68	0.68	0.68	0.68	0.68
No change							
Forest fires	324×10^6 T	2930 lb/T	0.39	0.39	0.39	0.39	0.39
No change							
TOTAL			14.08	15.88	20.59	28.20	40.57
RELATIVE CHANGE			100%	113%	146%	200%	288%

Revelle (57) points out that total recoverable fossil fuels as of 1965 are estimated to be about 10^{12} tons, with a CO_2 equivalent of 7.9×10^{18} g or 330% of the 1960 atmospheric CO_2. If, as indicated by the 1958–1963 data, half of this stays in the atmosphere, then ultimately the atmospheric CO_2 concentration could increase from 320 ppm to about 850 ppm, as recoverable fossil fuels are used up.

Most generalized analyses of the CO_2 situation infer that a given increase in atmospheric temperature could cause a gradual melting of the Polar Ice Caps. Revelle (57), for example, states that if half the energy associated with a 2% increase in radiation energy, as might occur from a 25% increase in CO_2, were available to melt the polar ice caps, the ice would disappear in about 400 years. Melting ice caps, if they occurred, would obviously result in inundation of coastal areas. However, when the details of the total atmospheric system are considered, it is not at all obvious what the result of added CO_2 might be.

Fletcher (76) is carrying out a long term detailed study of the relationship between the Antarctic Ice Cap and world climate. In Fletcher's research, consisting of developing models of ocean/atmosphere heat exchange in the Antarctic, he has found apparent global climatic relationships. Increases in ice area in the Antarctic were found to correspond to an intensification of the zonal circulation in the Southern hemisphere and to a lesser extent in the Northern hemisphere, accompanied by a warming trend over most of the northern high latitudes. Conversely, decreasing ice in the Antarctic, such as has been postulated as resulting from an increase in atmospheric CO_2, corresponds to a weakening of the zonal circulation in both hemispheres, an increased prevalance of meridianal circulation patterns in the Northern hemisphere, and a cooling over most of the northern latitudes (76).

These correlations by Fletcher are based on what has occurred in the past, and with our present knowledge we are not justified in predicting future effects of CO_2 on the basis of these correlations. There are too many complicating factors in the total ocean/atmosphere system to permit solution at this time. For example, Möller (75) points out that a fairly significant 10% increase in CO_2 can be counterbalanced by a 3% change in atmospheric water vapor or a 1% change in total average cloudiness. It is doubtful that we are equipped at present to detect changes of this magnitude in either water vapor or cloudiness.

Although, as we mentioned before, there are other possible sources for the additional CO_2 now being observed in the atmosphere, none seems to fit the presently observed situation as well as the fossil fuel emanation theory does.

F. Summary of Carbon Dioxide in the Atmosphere

Revelle and others have made the point that man is now engaged in a vast geophysical experiment with his environment, the earth. On the basis of our present knowledge, significant temperature changes could be expected to occur by the year 2000 as a result of increased CO_2 in the atmosphere. These could bring about long-term climatic changes.

More recently, McCormick and Ludwig (56) have provided an analysis of the possible worldwide change of atmospheric fine particles. An increase in fine particulate material may have the effect of increasing the reflectivity of the earth's atmosphere and to reduce the amount of radiation received from the sun. Thus, this effect would be the opposite of that caused by an increase in CO_2. The argument has been made that the large scale cooling trend observed in the Northern hemisphere since about 1955 is due to the disturbance of the radiation balance by fine particles and that this effect has already reversed any warming trend due to CO_2.

Any secular changes due to increases in pollutant aerosols would be limited almost exclusively to the Northern hemisphere. This would result from the relatively small amount of pollution in the Southern hemisphere and the slow or negligible transfer of aerosols from across the equator. The CO_2 cycle is such that concentrations in the two hemispheres are equal.

It is rather obvious that we are unsure as to what our long lived pollutants are doing to our environment. However, there seems to be no doubt that the potential damage to our environment could be severe. Whether one chooses the CO_2 warming theory as described by Revelle and others, or the newer cooling theory indicated by McCormick and Ludwig, the prospect for the future must be of serious concern.

It seems ironic that, in air pollution technology, we are so seriously concerned with small scale events, such as the photochemical reactions of trace concentrations of hydrocarbons and the effect on vegetation of a fraction of a part per million of SO_2, whereas the abundant pollutants (CO_2 and submicron particles) which we generally ignore because they have little local effect, may be the cause of serious worldwide environmental changes.

IV. ATMOSPHERIC CARBON MONOXIDE

A. Introduction

Carbon monoxide has been considered an important atmospheric pollutant for many years because of its prevalence in automobile exhaust and in the effluents from poor combustion. In spite of widespread knowledge about the toxicity of CO, there have been many fatalities resulting from exposure to excessive CO.   Because of the hazards it posed, initial air pollution studies of CO in urban atmospheres were carried out mostly to determine whether specific segments of the urban population, i.e., traffic police and tunnel toll takers, were exposed to toxic levels of CO.   However, in more recent urban atmospheric studies, such as the USPHS Continuous Air Monitoring Program, CO measurements have been included as an indication of urban air quality.

It has typically been assumed that the only sources of CO were combustion sources, and that CO presented a relatively simple pollutant situation. However, as we will describe later, some natural sources of CO now seem to be clearly indicated.   This is a new dimension in CO studies, and there are a number of new problems that must be included.   Of major importance is the nature of the scavenging mechanism for CO.   We will discuss various possibilities in some detail.

B. Atmospheric Carbon Monoxide Concentration

The presence of CO in urban atmospheres at concentrations as high as 50 to 100 ppm has been recognized for many years, but it was not discovered that CO was present in trace amounts in the ambient atmosphere until Migeotte (77) detected absorption lines in the solar spectrum around 4.7 μ. Later, Migeotte and Neven (78) confirmed the identification with measurements in Switzerland, and further spectra were also obtained by Shaw et al. (79).   Some confusion arose when Adel (80), using spectra obtained at Flagstaff, Arizona was unable to detect the CO lines found by Migeotte in his Ohio spectra.   However, when further studies in United States and Europe continued to show the CO lines in the solar spectra, it became generally

TABLE XI

Typical Atmospheric Carbon Monoxide Concentrations
Found in Nonurban Areas

Place	Date	Local time	Wind direction and velocity (mph)	CO concentration (ppm)
Camp Century, Greenland	7/3/65	1245	SE-8	0.90
Camp Century, Greenland	7/3/65	1245	SE-8	0.85
Camp Century, Greenland	7/5/65	0800	SE-15	0.24
Camp Century, Greenland	7/5/65	0805	SE-15	0.32
North Coast, California	6/23/65	1400	W-8	0.85
North Coast, California	6/24/65	1400	W-10	0.80
Coastal Forest, California	6/24/65	1130	Calm	0.80
Crater Lake, Oregon (7000 ft elev)	9/27/65	0905	NE-22	0.30
Crater Lake, Oregon	9/28/65	0835	Lt. & Var.	0.08
Crater Lake, Oregon	9/28/65	1650	W-4	0.06
Crater Lake, Oregon	9/29/65	0910	Lt. & Var.	0.03
Crater Lake, Oregon	9/29/65	1700	Lt. & Var.	0.05
Patrick Point, Calif., Coast	10/02/65	1630	S-5	0.04
Patrick Point, Calif., Coast	10/03/65	1800	W-10	0.04
Patrick Point, Calif., Coast	10/04/65	1145	W-10	0.04
Patrick Point, Calif., Coast	10/05/65	0840	E-2	0.80
Patrick Point, Calif., Coast	10/05/65	1320	W-5	0.06
Patrick Point, Calif., Coast	10/06/65	1320	Lt. & Var.	0.34
Patrick Point, Calif., Coast	10/06/65	1650	W-8	0.06

assumed that CO was a more or less permanent but perhaps highly variable atmospheric constituent. On the basis of the solar spectra from the early 1950s an average ambient CO concentration of 0.2 ppm was estimated (3).

In 1966, after the development of new instrumentation (81) ambient atmospheric CO data could be obtained from remote locations. Some of these early data are shown in Table XI. Subsequently a number of studies have been made in remote areas, including five shipboard sampling studies in the Pacific (82,83); measurements at Point Barrow, Alaska (84); and in Greenland (85). In addition, CO data have been obtained in the Atlantic area in 1968 (86), by Swinnerton et al. (87), and by Seiler and Junge (88). In these latter studies Swinnerton used a gas chromatographic analytical procedure, while Junge used an instrument similar to that designed by Robbins. These Pacific, Arctic, and Atlantic data were all sea level samples. However, Junge (89) has also made a limited number of airborne measurements, including several penetrations of the tropopause.

In the Pacific area important data were obtained on a voyage from San Francisco to New Zealand and on a round trip voyage between the United

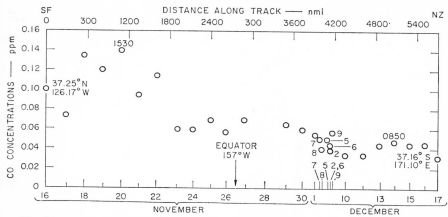

Fig. 10. Carbon monoxide concentration at local noon in Pacific Ocean, November–December 1967. Data taken along track from San Francisco to New Zealand during cruise 31 of USNS Eltanin.

States and Asia (82,83). Figure 10 shows the CO concentration at approximately local noon as measured on the San Francisco–New Zealand voyage (82). The general pattern shows increased CO concentrations from around 0.09 ppm near California to as high as 0.14 ppm East of Hawaii at 23° N, 140° W on November 20. From this point the decrease is relatively rapid to 0.06 ppm on November 23 at latitude of 12° N, 149° W. Concentrations remained at about 0.06 to 0.07 ppm until about December 1 at latitude 13° S, 172° W. Between December 1 and 10 there were variations between 0.04 and 0.06 ppm. After December 10 at 19° S, 176° W the average concentration was 0.04 ppm.

In June 1968 an East–West Pacific crossing was made on the USNS Perseus, with the return voyage in late July and early August, 1968. The average daily CO concentrations observed on these two trips are shown in Figure 11 (83). There does not seem to be any clear trend in these data, or at least nothing comparable with that shown for the North–South data in Figure 10. One possible significant point is that the data taken in July and August were higher than the June data. The average daily concentration for the westbound trip in June was 0.12 ppm, and for the mid-summer return trip the daily average was 0.18 ppm. In the Mid-Pacific generally East of the dateline, the East- and West-bound tracks of the Perseus coincided.

All of the available Pacific data have been combined to provide an indication of an average North–South CO concentration pattern over the Pacific. This is shown in Figure 12 (83). Average daily concentrations or individual

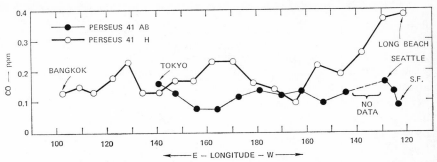

Fig. 11. Average daily carbon monoxide concentrations as a function of longitude across the North Pacific, USNS Perseus trips 41 A, B, H. June–July, 1968.

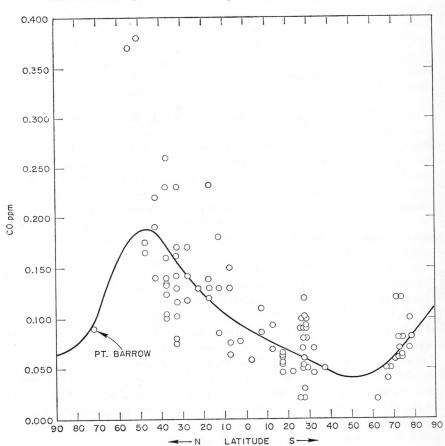

Fig. 12. Average latitude distribution of carbon monoxide in the Pacific area. Based on Stanford Research Institute data from USNS Eltanin and USNS Perseus voyages 1967–1968.

flask samples have been grouped into 5° latitude classes. The curve was fitted by eye and is an attempt to indicate an average North–South CO profile. Arctic concentrations in line with the measurements at Point Barrow, Alaska and in Greenland are about 0.1 ppm. The highest concentrations, about 0.2 ppm, are found between 30° and 50° N. Southward from 30° N, CO concentrations decrease to 0.09 ppm at the equator and then to about 0.04 at about 50° S. Then southward CO concentrations apparently increase to about 0.08 over Antarctica. This increase in the Antarctic is based on a series of 16 spot samples taken in January 1967 between 65° S and 77° S along 170° E (82). How persistent this feature might be is unknown, but it was clearly evident in this series of samples.

In the North Atlantic, data are much less regular than in the Pacific. This is due to the influence of contamination sources, especially in North America, and results in wide fluctuations in concentration data (86). Data from the subtropical Caribbean area were reported by Swinnerton et al. (87) as ranging between 0.2 and 0.3 ppm. Seilers and Junge (88) report similar data for the Canary Islands. These concentrations are somewhat higher than comparable data from the Pacific where concentrations would probably be at and below 0.2 ppm, but the differences are not large and probably not significant.

Recent data on CO background levels in Europe have been reported (88). Preliminary data show that during the influx of strong westerly winds over West Germany the CO concentration was never observed to drop below 0.14 ppm. In addition, some exploratory aircraft measurements in the upper troposphere over the European Arctic gave similar concentrations (89). Junge interprets these results as indicating a CO background level in the range of 0.1 to 0.2 ppm. These data are generally similar to the CO levels found in Greenland by Stanford Research Institute in the summer of 1967 (85) and to the Pacific data in latitudes of about 50° N shown in Figure 12. On several occasions these aircraft measurements (Seilers and Junge, 1969) obtained data from above the tropopause. In these situations CO concentrations dropped to very low levels of 0.03 ppm and lower. Thus it appears that CO is mainly a feature of the lower part of the atmosphere.

If the curve in Figure 12 is considered representative of general tropospheric CO concentrations, average hemispheric CO concentrations can be estimated. Integrating this average curve and correcting for the latitudinal change in atmospheric volume gives an average CO concentration in the Northern hemisphere of 0.14 ppm, 0.06 ppm in the Southern hemisphere, and a global average of 0.1 ppm.

Although our knowledge of ambient CO concentrations is still incomplete, an estimate of a global average concentration of 0.10 ppm, as indicated by the average Pacific data, seems reasonable. Using this average ambient CO concentration and an estimated emission rate permits calculation of the rate at which CO is removed from the atmosphere. At a CO concentration of 0.10 ppm, the mass of CO in the atmosphere is about 5.6×10^{14} g. Since the estimated annual total CO emission rate from pollutant sources alone is 2.5×10^{14} g, the atmosphere contains about 2 years' emissions. This is a surprisingly short time for CO, which as we shall see, has no clearly defined reaction cycle to provide a sink and where natural sources of CO also make important contributions to the atmospheric total.

A comparison of the solar spectra data from the 1950's which, as we indicated, averaged about 0.2 ppm shows that they are in agreement with the mid-latitude Stanford Research Institute chemical data where the observations were made, and does not give us any reason to believe that any major accumulation of CO has taken place in the approximately 15 years between these observations and the recent analyses. In 1968 Stanford Research Institute attempted to measure CO concentrations in air trapped in Greenland glacial ice and which could be dated back to about 1800. The expectation was that this would provide a way to determine long term average changes in atmospheric CO concentrations. The experiment was apparently unsuccessful; however, because high and generally uniform CO concentrations of about 0.3 ppm were obtained from the ice. It is speculated that the process of entrapment resulted in a gradual concentration of CO in the ice samples. This theory was substantiated by the measurement of CO concentrations in recent ice. Samples 10 to 40 years old all contained air with CO concentrations between 0.8 and 1.5 ppm. These concentrations are much higher than the current average concentrations over the ice cap. Gas chromatography was used to detect the CO and instrumental error was practically ruled out. It is believed that the results indicate qualitatively the presence of CO from natural sources in the atmosphere of 1800 and earlier, but no actual concentration level can be postulated (91).

C. Sources of Carbon Monoxide

The pollutant sources of CO have been recognized for many years and the amounts expected to be emitted to the atmosphere can be estimated with reasonable accuracy. This has been done in Table XII. On a worldwide basis the total annual CO emission from combustion sources is 304×10^6 tons or 2.8×10^{14} g.

The data shown in Table XII confirm the fact that automobiles are the major source of CO, since gasoline combustion accounts for 64% of the total. The magnitude of this source has been estimated, using a standard CO emission value of 2.91 pounds per gallon of gasoline (92). Annual worldwide gasoline production was 379×10^6 tons in 1966 (93) which, assuming no net carryover, provides a total CO emission of 193×10^6 tons.

TABLE XII
Estimated Annual CO Emissions

Source	Consumption	CO factor	CO emission, tons
Gasoline	379×10^6 T[a]	2.91 lb/gal[b]	193×10^6
Coal[c] (total)	3074×10^6 T[a]		
Power	1219×10^6 T[a,d]	0.5 lb/T[e]	
Industry	781×10^6 T[a]	3.0 lb/T[e]	12×10^6
Residential	404×10^6 T[a]	50.0 lb/T[e]	
Coke and gas plants, etc.	615×10^6 T	0.11 lb/T	
Wood and noncommercial fuel	1260×10^6 T[f]	70 lb/T[g]	44×10^6
Incineration	500×10^6 T	100 lb/T	25×10^6
Other industries[h]	—	—	30×10^6
			304×10^6 tons

[a] U.S. Stat. Abs. 1967 (93).
[b] Mayer 1965 (92).
[c] 13,100 Btu/lb U.S. Stat. Abs., 1967 (93).
[d] Nat. Soc. for Clean Air 1962–63 (95).
[e] Heller and Walters 1965 (96).
[f] U.N. 1963 (97).
[g] Darley et al. 1966 (98).
[h] HEW 1970 (99).

Coal production on a global basis amounted to 3089×10^6 tons in 1966 (93). The major CO emissions from coal were assigned to power development, industrial operations, including rail and steamship uses, residential and local uses, and coke and gas plants. Tonnages for the various uses of coal were obtained for United States and Great Britain (94). In Table XII the tonnage in each class is the sum of the United States tonnage plus that for the rest of the world. This latter figure assumes that the relative amounts in the various classes for areas outside the United States were the same as those which apply in Great Britain. Using this scheme the total annual CO emissions from coal total 12×10^6 tons.

The other major fuel is vegetal, both in the form of wood, bagasse, peat, and other noncommercial fuels. On the basis of 1952 data it was estimated that the annual use of wood fuel was 1260×10^6 tons (58) with a CO emission of 70 pounds per ton (98) for a total annual CO emission of 44×10^6 tons. Incineration emissions are estimated at 25×10^6 tons on the basis of the incineration of 500×10^6 tons per year and an emission factor of 100 pounds CO per ton.

A number of industries, especially petroleum refineries, gray iron foundries, and kraft paper mills, emit some CO in their operations (58). On a global basis, these sources annually contribute an estimated 30×10^6 tons of CO to the atmosphere.   Emissions of CO from fuel oil, natural gas, and diesel are not significant.

As shown previously, the total estimated annual CO emissions are 304×10^6 tons.   In 1952 Bates and Witherspoon (100) estimated CO emissions to be 210×10^6 tons annually.   Of this, 130×10^6 tons were attributed to automobiles, 40×10^4 tons to coal, and 40×10^6 tons to wood combustion and forest fires.   These changes seem reasonable considering the increased use of gasoline and the fact that coal is probably less of a source now because of the increase in the proportion that is being used in highly efficient power generation processes.

There is a strong imbalance in CO emissions between the Northern and Southern hemispheres.   On the basis of gasoline consumption, 95 % occurs in the Northern hemisphere (13).   Data on the other sources are not readily available, but it seems quite likely that the other combustion sources also show this strong bias toward the Northern hemisphere.   This is doubtless reflected in the differences in atmospheric concentrations that seem to be apparent.

Although natural sources of CO have not received much attention in the literature to date, there is increasing evidence that trace concentrations of CO in the atmosphere could be from natural sources.   Although there is no indication of CO being a significant portion of volcanic gases, the reported data are scarce and there are some references to the release of CO from heated basalt rock (101).   There is one report, by Wilks (102), that CO is given off by vegetation and, in particular, as a result of the action of ultraviolet radiation on chlorophyl.   As part of his experiment Wilks enclosed several growing branches in plastic bags.   After several days he measured relatively high CO concentrations in the air surrounding the growing branches, and he cited this as an indication of CO being produced by the plant.   It seems quite possible that the CO in this situation resulted from the artificial sealing off of the growing branches rather than from natural processes within the plant.   However, further study is certainly justified.   Some bacteria have been found that are capable of producing CO as a result of their metabolism (103).

The forest fire contribution can only be estimated roughly on the following basis.   First, in the United States in 1966, forest fires acreage was 4.5×10^6 acres (93); we assumed this was 25 % of the global forest fire areas, giving a world total of 18×10^6 acres burned annually.   Second, average fuel burned per acre was assumed to be 18 tons, which is roughly 90 % of the average amount of wood growing on an acre (13,104).   Third, average CO emission from bush and green wood was taken as 70 pounds per ton (98).   This procedure results in a total CO contribution of 6.3×10^6 tons annually.

This amount is clearly not a significant factor in the total annual emissions of CO, and more detailed refinements are not justified because they could hardly make a significant change in the total.

Important evidence of a natural CO source in the surface waters of the ocean has recently resulted from studies by Stanford Research Institute in the Pacific, and by the Naval Research Laboratory (NRL) in the Atlantic.

One indication of a significant natural marine CO source resulted from the study of the atmospheric measurements made in the Pacific in the fall of 1967 (82) during the oceanographic cruise from San Francisco to New Zealand. These data showed that CO concentrations had considerable diurnal variation. The only logical explanation for short period changes in CO in the middle of the Pacific seemed to be some action of the ocean either as a source or sink. The fact that CO concentrations in the Southern hemisphere, as shown by Figure 12, now seem to be one-half to one-third those in the mid-latitudes of the Northern hemisphere may indicate the importance of oceanic contributions in the Southern hemisphere.

Data from the Pacific in 1968 also seem to indicate, but do not prove, that the ocean could be a CO source (83). A weak diurnal cycle is indicated by average data, with lower concentrations occurring in early afternoon and higher concentrations at night. A seasonal change is also possible with August concentrations being higher than early June measurements.

More recently, Swinnerton and his coworkers (87,105) have shown that the surface water in the open ocean of the Caribbean was supersaturated relative to atmospheric CO concentrations. On a quantitative basis this supersaturation was as much as 90 times in one area and, in general, the ocean CO concentration was at least 10 times the equilibrium saturation. Atmospheric concentrations over the open ocean ranged from about 75 to 250 ppb, which is in the same range as Stanford Research Institute encountered in the Pacific. One obvious explanation of these data is to postulate a source of CO in the surface ocean water.

Although a source of CO in the ocean has not been identified, a logical explanation is that it is a biological process. The siphonophores may play a major role. For example, E. G. Barham (106) reports that siphonophores are consistently present in the deep-scattering layer of the ocean, and that sound is scattered where siphonophores are present in great abundance. Pickwell (107) investigated siphonophores and found that they repetitively expel bubbles containing nearly 80 % carbon monoxide. This deep-scattering layer is known to have a diurnal cycle, which is at the surface at night and descends to depths of 300 to 400 meters in the daytime. Pickwell calculated a possible normal and a maximum CO contribution to the atmosphere. The maximum contribution is 2 cc of CO per square meter of sea surface per day, with a more "normal" contribution of 0.02 cc of CO per square meter of sea surface per day.   This source of CO would emit 362×10^6 tons/year maximum, to 3.62×10^6 tons/year minimum.   The maximum contribution figure is based on unusually dense swarms of siphonophores observed only once with the maximum observed rate of bubble production.   The minimum figure is probably a more normal figure and is based on typical numbers of siphonophores seen or captured.   All production rates calculated by Pickwell assume no biochemical mechanism for removal of siphonophore-produced CO.   The diurnal cycle of the siphonophores bears an interesting correlation to the marine CO diurnal cycle observed by Stanford Research Institute on the Eltanin cruise from San Francisco to New Zealand.

Swinnerton et al. (108) have related the observed CO supersaturation of sea water to a possible oceanic CO source of about 10×10^6 tons/year.   This is comparable to the "normal" conditions estimated above for siphonophores.

Another potential natural source of CO over land areas follows from the observation by Went (109) that photoreactions of terpenes in the atmosphere are significant.   Went estimates that about 200×10^6 tons of volatile organics of plant origin are molecularly dispersed throughout the world each year.   Photochemical oxidation of these plant volatiles by ozone or nitrogen oxide could produce 12×10^6 tons of CO annually.   This calculation is based on the assumptions that one molecule of CO will be produced per three molecules of organics (110), and that the average organic molecular weight is 150.   A terpene source as large as 10^9 tons has also been estimated by Went (111); however the smaller value seems more realistic.

A basic question that now needs to be answered is whether the background CO in clean atmospheres originates from combustion or if biochemical utilization of CO_2 or carbonates produces CO.   The diurnal changes in CO concentration found by Stanford Research Institute in marine air of both the Northern and Southern hemispheres and the air/water saturation ratios found by NRL could indicate that the ocean, or biological processes within the ocean, could be a source of CO.   Data on remote land areas are not available, and thus no direct calculation can be made of the participation of land plants in a natural CO cycle.

One other presently unexplained link between CO and the biosphere exists in the case of *Nereocyctis*, the familiar kelp or seaweed with large floats or bladders.   The floats of this kelp have been found to contain CO at concentrations of up to 800 ppm.   This is obviously much higher than exists in the atmosphere and probably also in the ocean.   The question as to how this occurs, whether it is just the unused residue from material absorbed from the water or whether it was produced by processes within the plant, has not been resolved.

Our current estimates of CO emissions from both urban and natural sources are summarized in Table XIII.   The total emission is 337×10^6

TABLE XIII
Estimated Total Emissions of Carbon Monoxide
from Urban and Natural Sources

Source	Annual emission (tons/year)
Gasoline	193×10^6
Coal	12×10^6
Other industrial	30×10^6
Noncommercial fuel	44×10^6
Incineration	25×10^6
Forest fires	11×10^6
Marine siphonophores	10×10^6
Terpene reactions	12×10^6
TOTAL	337×10^6

tons/year, considering a low value for oceanic emissions. Of this total, 33×10^6 tons should be attributed to natural sources, including forest fires; and the remainder or 88% is attributed to pollutant sources.

D. Scavenging Processes for Atmospheric Carbon Monoxide

Carbon monoxide at the present time is an important air pollution enigma because, in spite of its ubiquitous presence in combustion gases, we cannot enumerate any known scavenging mechanisms. That there must be one or more fairly effective processes by which CO is removed from the atmosphere is shown by the fact that, as we have mentioned, the present emission rate would double present atmospheric concentrations in about 3 years. There is no evidence for any large scale increase of atmospheric CO concentrations, and thus we must conclude that yet-to-be-proven scavenging processes exist in our environment.

Several possible scavenging mechanisms can be postulated (gaseous reactions, absorption processes, and biological processes). These various possible processes will be evaluated briefly in the following sections.

1. Gas Phase Reactions

Bates and Witherspoon (100) reviewed various likely gas phase reactions whereby CO might be oxidized to CO_2. The reaction with molecular oxygen

$$2CO + O_2 \rightarrow 2CO_2$$

does not occur at ordinary atmospheric temperatures. Mellor (112) reports an experiment in which CO and O_2 mixtures remain unchanged after 7 years' exposure to sunlight. The possibilities of CO reactions with atomic oxygen have been reviewed by Leighton (110) and he concludes that they are unimportant.   Bates and Witherspoon (100) came to a similar conclusion. Ozone will oxidize CO to CO_2, but according to Leighton, this reaction is extremely slow at atmospheric temperatures and concentrations. A high activation energy of about 20 kcal for the CO oxidation by O_3 was found by Zatsiorskii et al. (113).   Oxidation by NO_2 in the reaction

$$NO_2 + CO \rightarrow CO_2 + NO$$

has an even higher activation energy of about 28 kcal (114), which essentially rules out this reaction.

There has also been some consideration given to the possibility that some very fast reactions occur between CO and certain intermediates of photochemical smog reactions.   One possible intermediate is the hydroxyl radical. It can be produced, for example, by the photolysis of aldehydes to produce perhydroxyl, which can then be reduced to the hydroxyl radical.   According to the studies of Greiner (115) hydroxyl is a very rapid reactant with CO and there is some indication that perhydroxyl also reacts with CO.   These reactions are chain-type, beginning with CO and hydroxyl which produces CO_2 and H atom, the H atom oxidizes to perhydroxyl, which in turn will oxidize CO to CO_2, reforming hydroxyl.   Doyle (116) calculates that a global average hydroxyl concentration of 10^{-9} to 10^{-8} ppm would be sufficient to convert all emitted CO to CO_2.   The significance of this hydroxyl sink mechanism remains to be proven.   A possibly serious limitation to this mechanism is the potential effect of methane (CH_4), which can also be oxidized by hydroxyl.   Since, throughout most of the troposphere, CH_4 is an order of magnitude higher in concentration than CO, 1.5 ppm versus 0.1 ppm, the major reaction for available hydroxyl may be primarily with CH_4.   Only detailed and careful research will provide an answer.

Although future research with hydroxyl class radicals may be rewarding, at present we must conclude that there are no proven significant gaseous oxidation reactions for CO in the ambient atmosphere.

2. Absorption Processes

Absorption of CO on surfaces exposed to the atmosphere does not represent a sink unless some reaction can be postulated that would promote the oxidation of CO on the surface.   No such reaction has been discovered.

Oceans are a recognized sink for atmospheric CO_2, and it is natural to ask whether they are also a sink for CO.   Although CO solubility in seawater is about 20 ml per liter (117), at present there is no reason to believe that the oceans are a CO sink because there is no recognized process or reaction that would remove CO from solution.   This is in contrast to the various physical and biological processes that CO_2 can undergo.   The solubility of CO is not sufficient without some reaction, because if the concentration of CO in the ocean follows Henry's law and is proportional to the partial pressure of CO in the atmosphere, a typical remote area, ambient concentration of 0.10 ppm results in a possible total of only 3.4×10^{12} g of CO in solution in the oceans of the world (volume $= 1.37 \times 10^{21}$ liters) even assuming uniform mixing. This is only about 2% of the estimated yearly production of CO. This clearly eliminates the ocean as a CO sink unless there are reactions occurring to remove CO from solution. The previously mentioned diurnal cycle for CO concentrations over ocean areas and the supersaturation of sea water both seem to indicate some participation of the ocean in the CO cycle. The high CO concentrations, up to 93%, in some marine plants (117) also indicate some complex, probably biological, system involving CO in the ocean. Because of low CO solubility and no available reaction mechanism, rain washout cannot be a significant CO removal mechanism.

3. Biological Processes

Even though no sink mechanism for atmospheric CO has been identified in atmospheric and surface reactions or absorption, the present atmospheric concentrations and the total emissions clearly indicate that some mechanism is removing large quantities of CO from the atmosphere. One other possible sink mechanism for atmospheric gases such as CO is in both the terrestrial and the marine biospheres.

At least two types of CO scavenging processes can be postulated for the terrestrial and marine biological populations. One is related to the presence in significant numbers of specific species of plants or animals that could metabolize CO. A second mechanism is a biochemical one in which CO would be bound to widely distributed organic compounds in much the same way it is bound by hemoglobin in blood.

One obvious biological sink of the metabolizing type could be bacteria-promoted reactions that use CO as a source of carbon, i.e., *Bacillus oligocarbophilus* (103). Jaffe (118) has summarized other possible scavenging mechanisms involving microorganisms, including some of the methane-producing soil bacteria. No estimate can be made at this time as to the distribution of bacteria of this type either on land or in the oceans, and therefore the practicality of regarding bacteria as a significant CO sink is unknown. However, laboratory studies in small exposure chambers have been reported by Levy (119) in which CO rapidly disappeared, presumably as a result of soil organisms. These experiments are some of the first that indicate a CO scavenging process that could be of sufficient magnitude to maintain current pollutant emissions in a quasi-equilibrium state. Another potential biochemical scavenging process for CO is the binding of CO to the porphyrin type compounds that are universally distributed in living materials. In particular, the heme compounds, which contain cross-linked iron, could conceivably bind CO, at least on a temporary basis and at a rate that is a function of atmospheric concentration. Since porphyrin compounds are so widely distributed, this type of process could perhaps have great potential for scavenging CO. Permanent removal from the environment would depend on whether CO subsequently reacted to form CO_2 when the porphyrin compound is broken down. Plant systems could provide a site for such reactions.

Within the biosphere, the process of plant respiration may also provide potential CO scavenging. That CO does react with plants is shown by the fact that a number of effects have been noted in plants after exposure to various concentrations of CO. One specific reaction is that between CO and cytochrome-oxidase. A number of different effects resulting from CO exposures are mentioned by Ducet and Rosenberg (120). The most frequent effect due to CO seems to be the inhibiting of plant respiration processes. Burris (121) describes the inhibiting effect that CO has on nitrogen fixation. Exposure to relatively high concentrations, i.e., 0.01% to 1.0%, has been reported by Carr (122) to cause a variety of visible effects.

With these various effects of CO on plants having been noted, it is perhaps logical to conclude that CO entering the plant during respiration can participate in reactions within the plant and that these plant processes would provide a means for the removal of CO from the atmosphere.

The possibility that CO is produced rather than scavenged by vegetation must also be recognized and, as with the marine system, the only evidence to date indicates that CO is produced by vegetation. This is based on the fact that a limited number of tests by Wilks (102) seemed to show CO production when he completely enclosed some growing branches for several days.

4. Summary of Scavenging Processes

In the previous discussion we have described only tropospheric reactions and have ignored the photochemical reactions in which CO can participate in the high stratosphere or mesosphere. If these processes were an important CO sink, the atmospheric residence time for CO would have to be 50 to 100 years because of the slow vertical transport. We believe now that the residence time of CO is in the range of one to 3 years, and thus the active scavenging mechanisms must be present at the earth's surface or at least within the troposphere.

Of the various scavenging processes presented in this discussion, none has been proven to be the major process responsible for the scavenging of CO from the atmosphere. However, the research on soil organisms seems to provide a reasonable explanation and more definitive analyses can be expected in the near future.

V. ATMOSPHERIC NITROGEN COMPOUNDS

A. Introduction

The total circulation of nitrogen compounds through the earth's environment is not well known, and because of the many different compounds involved the nitrogen circulation is obviously a complex one. Nitrogen accounts for roughly 80%, or 4.6×10^{17} tons, of the earth's atmosphere; it is relatively inert and not especially soluble in water. Under high temperature conditions and in various biological reactions various oxides of nitrogen are formed. Nitrogen is essential for plant growth and nitrogen is fixed from the atmosphere by some plants. The breakdown of various organic nitrogen compounds produces ammonia and possibly some of the oxides.

Although there are eight different oxides of nitrogen, only three have been shown to be of any importance in the atmosphere. These are: N_2O, nitrous oxide; NO, nitric oxide; and NO_2, nitrogen dioxide. All are gases at normal conditions. Another important atmospheric gaseous nitrogen compound is ammonia, NH_3. Nitrogen compounds also exist as aerosols of HNO_3, nitric acid; NO_2^-, nitrite; NO_3^-, nitrate;* and NH_4^+, ammonium.*

Of these seven forms of atmospheric nitrogen, the only significant pollutants emitted by man's activities are NO and NO_2. On a total mass basis, however, pollutant emissions account for a very minor fraction of the total circulation of nitrogen compounds, as will be seen.

Applicable atmospheric information on these nitrogen compounds in the ambient atmosphere is not voluminous and is to a large extent oriented toward high concentrations in polluted urban areas. Some detailed studies of NH_3 were carried out in the late nineteenth century by biologists. These data now appear to be rather imprecise, and very little recent work has been done to upgrade our knowledge of NH_3.

B. Sources of Atmospheric Nitrogen Compounds

1. Nitrous Oxide

The most plentiful atmospheric N compound is the relatively inert N_2O which has a mean concentration of about 0.25 ppm. The major source for N_2O is apparently biological action in the soil, the N_2O being evolved from the decomposition of nitrogen compounds by soil bacteria (123). Arnold (124) on the basis of detailed experiments of bacterial action on NH_4 and NO_3 found that N_2O and some N_2 were produced. The reaction conditions were

---

* In subsequent remarks we will use NO_3 when referring to nitrates and NH_4 for ammonium.

complex, however, depending on soil aeration and other factors.   Junge (3) has summarized the research of Arnold (124) as well as of Goody and Walshaw (123) and has concluded that the total amount of fixed N in the soil can be transformed by bacterial action in a period of between 100 and 1000 days, depending on the assumptions used.   This is also the conclusion of Georgii (27).

Another possible source or sink is the ocean.   Data of Craig and Gordon (125) show that the amounts of dissolved N_2O to depths of 5000 meters in the South Pacific are slightly lower than the equilibrium concentration for atmospheric N_2O.   These results, however, can only be considered very preliminary because of the accuracy of the methods used.   If confirmed, these results would show that the ocean may be a sink but a very inefficient one since the vertical concentration gradients in the ocean water are apparently rather small.

The possibility of photochemical production of N_2O in the stratosphere from the reaction

$$N_2O_3 \rightarrow N_2O + O_2$$

was put forth by Bates and Witherspoon (100) as a possible source mechanism to balance the recognized photochemical destruction of N_2O above altitudes of 30 km.   However, experiments by Goody and Walshaw (123) showed that this reaction was apparently a minor N_2O source and could balance only about $2\frac{1}{2}\%$ of the probable photochemical destruction processes.

On the basis of their calculation of the turnover time of fixed N in the soil, Goody and Walshaw (123) concluded that an average worldwide production rate of 8×10^{10} molecules/cm^2 sec (10^9 tons/year) was a reasonable expectation.   This neglects any possible production in the ocean and is the best estimate presently available for N_2O production.

2. Nitric Oxide and Nitrogen Dioxide

There is qualitative indication that NO can be formed by bacterial action in the reduction of N compounds under anaerobic conditions.   The NO is then rapidly oxidized in the air to NO_2.   For example, production of NO_2 in very high concentrations, several hundred parts per million, has been noted in closed silos containing a number of different materials, causing some very hazardous conditions for farm workers (126,127).   This NO_2 formation process has been postulated to be one in which the nitrates in various materials are first reduced to nitrites by bacteria (128).   This step is followed by the conversion of the nitrites to nitrous acid, HNO_2, which in turn is broken down to NO.   It is not known whether this process can proceed at a significant rate in soil and under conditions quite different from those found in a closed silo.   Junge (3) points out that the formation of

NO from HNO_2 occurs in soils which are acid and mentions this as a reason for acid soils characteristically to have a negative nitrogen balance. However, this natural emission of NO is considered to be the best explanation for the rural area measurements of Ripperton et al. (129), Hamilton et al. (130), and Lodge and Pate (131).

The fixation of N by lightning has been investigated in a number of studies. Georgii (27) concluded that lightning was unimportant in the fixation of N, and according to Junge (3) the consensus is that the evidence of lightning fixation of N is marginal at best.

A variety of upper atmospheric reactions involving both NO and NO_2 along with atomic N and O are listed by Altshuller (126). For example, twilight and night air glow are mentioned as possible indications of one or more of these reactions. However, it seems unlikely that these processes are significant in the nitrogen cycle in the lower atmosphere.

The pollution sources of NO and NO_2 will be considered together and expressed as NO_2. This is necessary because the available data on emissions seldom distinguish between the two.

The pollution sources of NO and NO_2 are primarily combustion processes in which the temperatures are high enough to fix the N in the air and the combustion gases are quenched rapidly enough to reduce the subsequent decomposition back to N_2 and O_2.

An estimate of worldwide NO_2 emissions has been made from estimates of fuel combustion processes from various sources and from NO_2 production ratios (92). Our estimate of NO_2 emissions is shown in Table XIV. As shown, the estimated annual production is about 53×10^6 tons with 51 % of the total due to coal combustion and 41 % due to petroleum production and the combustion of petroleum products. Natural gas is a relatively minor source, 4 %, and the combustion of wood is an insignificant source. This 53×10^6 tons is probably a minimum value for total NO_2 production because of possible sources in the soil which cannot be quantitatively estimated.

3. Ammonia

Ammonia is a relatively unimportant industrial emission, but in the overall atmospheric nitrogen cycle NH_3 plays a very significant role. The major sources of NH_3 are found in the biosphere. Junge (3) points out that land and ocean areas can be considered sources for biologically generated NH_3, although some measurements in both environments indicate that under certain conditions these areas can also act as sinks. In soils the release of NH_3 is apparently a function of pH, and alkaline soils favor the release of NH_2. The determination of the release rate is complicated by the many soil condition factors that affect it. Both Junge (3) and Altshuller (126) indicate

TABLE XIV

Worldwide Urban Emissions of Nitrogen Oxides
(as NO_2)

Fuel	Source type	Fuel usage	Emission factor	NO_2 emission
Coal	Power generation	1219×10^6 tons[a]	20 lb/ton[d]	12.2×10^6 tons
	Industrial	1369×10^6 tons[a]	20 lb/ton[d]	13.7×10^6
	Domestic/Commercial	404×10^6 tons[a]	5 lb/ton[d]	1.0×10^6
Petroleum	Refinery production	$11,317 \times 10^6$ bbl[a]	6 ton/10^5 bbl[e]	0.7×10^6 tons
	Gasoline	379×10^6 tons[a]	0.113 lb/gal[d]	7.5×10^6
	Kerosene	100×10^6 tons[a]	0.072 lb/gal[d]	1.3×10^6
	Fuel oil	287×10^6 tons[a]	0.072 lb/gal[d]	3.6×10^6
	Residual oil	507×10^6 tons[a]	0.104 lb/gal[d]	9.2×10^6
Natural gas	Power generation	2.98×10^{12} ft³[b]	390 lb/10^6 ft³[d]	0.6×10^6 tons
	Industrial	10.27×10^{12} ft³[b]	214 lb/10^6 ft³[d]	1.1×10^6
	Domestic/Commercial	6.86×10^{12} ft³[b]	116 lb/10^6 ft³[a]	0.4×10^6
Others	Incineration	500×10^6 tons	2 lb/ton[d]	0.5×10^6 tons
	Wood fuel	466×10^6 tons[c]	1.5 lb/ton[d]	0.3×10^6
	Forest fire	324×10^6 tons[a]	5 lb/ton[f]	0.8×10^6
TOTAL				52.9×10^6 tons

[a] 1967 U.S. Statistical Abstracts (93).

[b] Figure is 1.28 × U.S. usage as per 1967 U.S. Statistical Abstracts (93).

[c] World Forest Inventory, U.N. 1963 (97).

[d] Mayer, M., "Pollutant Emission Factors," USPHS, May 1965 (92).

[e] Elkin, H. F, In: "Air Pollution," Vol. II, A. C. Stern, Ed., Academic Press, N.Y. 1962 (132).

[f] Gerstle, R. W. and D. A. Kemnitz, J. Air Pollut. Control Assoc. **17**, 324 1967 (133).

that the main biological source of NH_3 is the bacterial breakdown of amino acids in organic waste material. Various mechanisms have been described by Gorle (136), and Allison (137) has reviewed some of the complexities of NH_3 release from the soil.

Although urban emissions of NH_3 are relatively unimportant, there are some which can be catalogued. Ammonia emissions are primarily from combustion sources, but some specific industrial processes can also be minor contributors. Most emission surveys neglect NH_3 because except in breakdown situations and unless an odor is apparent there is no known harmful effect (139). Summaries of urban and industrial emission data prepared by Wohlers and Bell (134) and by Wohlers (139) list emissions for NH_3 on the basis of fuel combustion. These factors have been used to estimate the likely worldwide industrial emissions of NH_3 given in Table XV. The total tonnage of NH_3 from these sources, 4.2×10^6 tons, is significantly less than the previously calculated emissions for NO_2, SO_2, or CO.

No measure of soil or oceanic emission rates or factors for NH_3 have been discovered. However, the nitrogen-containing particulate matter deposited from the atmosphere has been measured and identified by Hoering (140) as originating from NH_3, either in solution as NH_4, or by oxidation of NH_3 to nitrite and nitrate. This reaction process of NH_3 to NO_3 has often been considered a necessary one to explain the observed situation. However, as

TABLE XV

Worldwide Urban Emissions of Ammonia

Fuel	Source	Fuel usage	Emission factor[a]	NH_3 emission, tons
Coal	Power generation	1219×10^6 T[b]	2 lb/T	1.2×10^6
	Industrial and others	1773×10^6 T[b]	2 lb/T	1.8×10^6
Petroleum	Refinery catalytic cracking operations	2410×10^6 bbl[c]	0.7 T/10^5 bbl	0.02×10^6
	Fuel oil	794×10^6 T[b]	2 lb/T	0.8×10^6
Natural gas	All uses	20.56×10^{12} ft[3][d]	0.010 T/10^6 ft[3]	0.2×10^6
Others	Incineration	500×10^6 T	0.015 T/100 T	0.08×10^6
	Wood	466×10^6 T[e]	0.12 T/100 T	0.06×10^6
	Forest fires	324×10^6 T[b]	0.015 T/100 T	0.05×10^6
				4.21×10^6

[a] Wohlers and Bell 1956 (134).

[b] U.S. Stat. Abs. 1967 (93).

[c] Includes estimate for USSR operations, Uhl, 1965 (135).

[d] Value is 1.28 times U.S. usage as per U.S. Stat. Abs. 1967 (93).

[e] U.N. 1963 (97).

we will point out in a subsequent section, we have found no oxidation process by which the NH_3 could be converted to nitrite in the atmosphere. In addition, as shown in the nitrogen cycle presented in this chapter, it does not seem necessary to have a conversion of NH_3 to NO_3.

4. Nitrite, Nitrate, and Ammonium Compounds

Nitrogen is found in the atmosphere in aerosol form as NO_2^-, NO_3^-, and NH_4^+ compounds. As indicated in the next section, these materials can result from the atmospheric reactions of both NH_3 and NO_2. The most likely scavenging reaction for aerosols is through solution in fog, cloud, and rain droplets. In precipitation NH_4 is the most prevalent of these compounds. Eriksson (141) estimated that approximately 75% of the total nitrogen in precipitation appears as NH_4. In our previous discussion of sulfur in the atmosphere it was pointed out that if NH_3 was present in fog or cloud droplets that SO_2 was rapidly oxidized in the droplet to form $(NH_4)_2SO_4$. Studies of both tropospheric and stratospheric aerosols have indicated that $(NH_4)_2SO_4$ is one compound which is typically found in most aerosol samplings (3). Thus these nitrogen-containing salts are important only as end products of the various gases emitted to the atmosphere.

There are, without doubt, some urban emissions of these particulate nitrogenous compounds from specific industrial processes, but outside the local source area their importance seems insignificant.

5. Organic Nitrogen Compounds

The organic nitrogen compounds which are of major concern in air pollution are of the peroxyacyl nitrate (PAN) family (138). These are the reaction products of photochemical reactions in the atmosphere involving hydrocarbons and NO_2. They will be discussed further in our discussion of hydrocarbons in the atmosphere. The formation of compounds in the PAN family is a keystone in understanding photochemical air pollution reactions. However, these are transient materials in the conversion of NO and NO_2 to an NO_3 aerosol, and outside the urban pollution source areas the NO_3 aerosol is the material which will be the longer lived atmospheric component. As we have seen, however, the fraction of the total nitrogen compound aerosols in the atmosphere which can be attributed to urban pollution is very small.

C. Atmospheric Conversion Reactions of Nitrogen Compounds

1. Nitrous Oxide

Nitrous oxide is essentially an inert gas at normal temperatures and pressures. Therefore, it takes no part in the tropospheric reactions involving nitrogen compounds. Bates and Hays (142) and Bates and Witherspoon

(100) have examined the photodissociation which can take place in the stratosphere and indicate that the processes

$$N_2O + h\nu \rightarrow N_2 + O(^1D)$$

and

$$N_2O + h\nu \rightarrow NO + N(^4S)$$

are the most significant ones. The latter process occurs at wavelengths below about 2500 Å. Bates and Hays (142) have indicated that the latter process, dissociation of N_2O into NO and atomic nitrogen, accounts for about 20% of the dissociation in the stratosphere. Their calculations, in which the rate is considerably less than has been previously calculated, show a dissociation rate of about 3×10^9 molecules/cm^2 sec or 3.5×10^{13} g/year, worldwide. This seems to be the only likely N_2O reaction.

2. Nitric Oxide and Nitrogen Dioxide

Nitric oxide and nitrogen dioxide are the only significant pollutants of the nitrogen compounds. Their importance as pollutants arises mainly from their participation in photochemical reactions involving reactive organics and SO_2. This has already been mentioned briefly in our discussions of SO_2 reactions. The variety of reactions which NO and NO_2 can undergo in polluted atmospheres has been covered in detail by Leighton (110).

The reaction of NO with O_3 in the atmosphere to form NO_2 is rapid and since there is always some background O_3 from stratospheric transport it has been generally assumed that NO_2 rather than NO is the predominant of the two species found in clean atmospheres. However, this assumption may have to be reversed on the basis of the findings of Lodge and Pate (131) and Ripperton and co-workers (129,130) that there are not only large concentrations but actually higher concentrations of NO than NO_2 in remote areas.

The O_3 oxidation of NO_2 to N_2O_5 is about 500 times slower than the O_3 oxidation of NO. Nevertheless, the oxidation of NO_2 is rapid enough to limit the half-life of 1 ppb NO_2 in the atmosphere in the presence of 5 ppb O_3 to about 2 weeks. The residence time of NO_2 based on our atmospheric nitrogen cycle is only 3 days and this short time must be explained by the scavenging reaction for NO_2

$$3NO_2(v) + H_2O(v) \rightleftarrows 2HNO_3(v) + NO(v).$$

The equilibrium constant is 0.004 atm^{-1} at 25° C (143), but with the extreme excess of water vapor found in the atmosphere 10% of the NO_2 is converted to HNO_3 at equilibrium. This HNO_3 vapor is rapidly removed by reaction with atmospheric ammonia and absorption by hygroscopic particles. At relative humidities higher than 98%, condensation of dilute nitric acid droplets will occur at $10°$ C and a HNO_3 vapor concentration of 0.1 ppb. All the HNO_3 eventually becomes nitrate salt aerosol.

The reactions of NO and NO_2 with olefins will be discussed briefly in Section VI.

3. Ammonia

We have shown that the biosphere must be a vast source of atmospheric ammonia. Measurements of atmospheric concentrations of NH_3 in various parts of the world indicate that both the land and the ocean segments of the biosphere are source regions. Comparison of the atmospheric concentration measurements of several investigators (3,27,130,141), indicates that background NH_3 concentration in the lower troposphere is latitude dependent, rising from about 8 ppb in mid-latitudes to over 20 ppb near the equator.

The total atmospheric NH_3 burden, which is produced by the bacterial decomposition of organic material and the burning of coal, is removed via several routes. Nearly three-fourths of the NH_3 is converted to ammonium ion condensed in droplets or particles. The reaction can occur as a direct NH_3 vapor-acid aerosol neutralization to form $(NH_4)_2SO_4$ or NH_4NO_3 or as a stepwise hydrolysis to form condensed NH_4^+ ion followed by neutralization. These reactions are known to be rapid and their velocities increase with increasing water vapor concentration. Thus, although the kinetics are complicated by the great variation in water content, the chemistry and mechanisms appear to be straightforward and simple.

The reaction of NH_3 with O_3 is extremely slow. Photooxidation can only occur high in the upper atmosphere (70 km or higher) and must be inconsequential in the troposphere. The direct gas-phase reaction between NH_3 and NO_2 is essentially a neutralization reaction producing a condensed particle phase. Thus, the usual strong oxidizers in the atmosphere can have no role in the oxidation of NH_3.

These aerosols result from reactions involving NH_3 and NO_2. As aerosols they follow the same reaction processes as previously described for SO_4 particles, i.e., coagulation, washout, rainout, and "dry" deposition. There seems to be no reason to expect different behavior of these materials because of their chemistry and so no further details will be given here. The reader is referred to Section III and our discussion of SO_4 aerosol reactions and scavenging processes.

4. Discussion of Nitrogen Reactions

The reactions involving N_2O, NO_2, and NH_3, which are of the most importance are those which provide for the formation of NH_4 and NO_3 particulate material. Processes by which NH_3 is scavenged by oxidation to NO_3 are mentioned as significant ones for atmospheric reactions but they have not been defined (144).   In fact, it is quite possible that this may be an unlikely atmospheric reaction.   Furthermore, NH_3 oxidation to NO_3 does not seem to be a necessary phase of the environmental nitrogen cycle discussed later.

D. Atmospheric Concentrations

1. Nitrous Oxide

For the several gaseous and particulate nitrogen compounds (N_2O, NO, NO_2, NH_3, NO_2^-, NO_3^-, and H_4^+) the only very satisfactory data on ambient atmospheric concentrations are those available for N_2O.   This is due to at least two facts.   First, since N_2O is practically inert it provides a means of studying large-scale transport in the atmosphere.   Second, photochemical reactions in the upper atmosphere involving N_2O are of interest in atmospheric physics.   Thus a good degree of effort has been spent on determining the concentrations, sources, and sinks of N_2O.   In addition, the fact that there are no pollutant sources of N_2O has meant that conditions in the ambient atmosphere are not complicated by diffusion patterns from large local sources.

Although early N_2O data tended to indicate concentrations averaging about 0.5 ppm; in recent years more sensitive measuring equipment has reduced this estimate by about 50%.   The best estimate now is that N_2O in the troposphere has an average concentration of about 0.25 ppm (3,142).

The Gutenberg University at Mainz has conducted an N_2O measurement program, and daily measurements over a period of about one year indicate in general a smaller variability than the previously published data, which were primarily only a short series of measurements (6).   In the newer data, variations from day to day as well as over periods of weeks are apparent. However, these are difficult to correlate with meteorological phenomena. A few aircraft measurements up to tropopause level show a rather uniform N_2O mixing ratio.   This confirms the older profile results of Adel (145), Goody and Walshaw (123), and Goldberg and Mueller (146).

2. Nitric Oxide and Nitrogen Dioxide

In contrast to the N_2O data, information on the ambient concentration of NO and NO_2 in the ambient atmosphere is scarce, and our understanding of these gases in areas beyond the pollution envelopes of our urban areas is very imperfect.   Table XVI summarizes the available data on NO and NO_2 in remote areas.   In Panama with fresh Caribbean air Lodge and Pate (131) found NO_2 concentrations in the dry season to range from less than 0.5 to 1.4 ppb with an average of 0.9 ppb.   Concentrations in the Caribbean air in the rainy season increased, with an average of 3.6 ppb and a range of 2.6 to 5.0 ppb.   Within the Panama forest NO_2 concentrations ranged from less

TABLE XVI
Summary of Nitrogen Oxide Concentrations

Location	Results		Reference
Panama			Lodge and Pate, 1966 (131)
(Marine)	NO_2:	0.5–5 ppb	
(Forest)	NO_2:	0.5–4 ppb	
	NO:	0 –6 ppb	
Mid-Pacific	NO_2:	94% of samples < 1 ppb	Lodge et al., 1960 (148)
Florida	NO_2:	0.9 ppb	Junge, 1956 (147)
Hawaii	NO_2:	1.3 ppb	Junge, 1956 (147)
Ireland	NO_2:	0.34 ppb (Spring)	O'Connor, 1962 (149)
	NO_2:	0.2 ppb (Fall)	
North Carolina	NO_2:	4.0 ppb	Ripperton et al., 1968 (129)
	NO:	2.6 ppb	
Pike's Peak	NO_2:	4.1 ppb	Hamilton et al., 1968 (130)
	NO:	2.7 ppb	
Antarctic	NO_2:	Avg < 0.6 ppb	Fischer et al., 1968 (150)

than 0.5 to 4.0 ppb while downwind from the forest area 3 samples from 20 miles offshore in Panama Bay averaged 1.6 ppb.

A few NO samples were also taken by Lodge and Pate (131) and concentrations of up to 6 ppb were found. The samples were insufficient for a meaningful average. However, the identification of NO as a constituent of the Panama atmosphere was quite significant because it was apparently the first indication that NO was a possible trace constituent of the ambient atmospheres.

Prior to these Panama samplings Junge (147) had reported an average NO_2 concentration of 0.9 ppb for 13 samples taken in Florida in 1954, an average of 1.3 ppb for 14 samples taken from the east coast of Hawaii, and an average of about 1.0 ppb for 4 samples taken at an altitude of about 10,000 feet on Mauna Kea.

Data on NO_2 concentrations in the mid-Pacific were obtained by Lodge and co-workers in 1957 and 1958 (148). Their limit of detection for this series was 1 ppb and in their approximately 2 months of sampling 94% of the NO_2 samples were less than 1 ppb. Only one value in this series exceeded 5 ppb.

O'Connor (149) reported some spring and fall NO_2 data from Mace Head, a remote coastal location in County Galway, Ireland. In April, concentrations ranged from 0.08 to 1.0 ppb with an average of 0.34 ppb from 9 observations. In September and October, 8 observations ranged from 0.1

to 0.34 ppb and averaged 0.2 ppb. These data are much lower than those reported by either Junge or Lodge and Pate. From his data, O'Connor concluded that there is an NO_2 background concentration which comes from some kind of natural source, rather than from any pollution activities.

In the area of geophysical levels of NO and NO_2, the work of Ripperton (129) and Hamilton (130) and their colleagues is probably the most significant. The data collected both in North Carolina and on Pike's Peak, Colorado, showed that NO as well as NO_2 was an important trace constituent of the atmosphere and, as shown by Table XVI, the concentrations of these two gases could be in the range of 1 to 5 ppb. These are the same general levels as measured by Lodge and Pate (131) in Panama.

On the basis of the data shown in Table XVI we have assumed the following mean conditions for our atmospheric cycle estimates for nitrogen compounds.

Land areas between 65° N and 65° S:	NO = 2 ppb
	NO_2 = 4 ppb
Land areas north and south of 65°	NO = 0.2 ppb
and all ocean areas:	NO_2 = 0.5 ppb

Applying these values to the various global sectors results in an estimated total atmospheric mass for NO of 4.7×10^6 tons (4.2×10^{12} g) and for NO_2, 13×10^6 tons (11.8×10^{12} g). On a nitrogen basis the totals are 2.2×10^6 tons (2×10^{12} g) of NO–N and 4.0×10^6 tons (3.6×10^{12} g) of NO_2–N.

3. Ammonia

Concentrations of NH_3 have generally been studied by the same investigators as those mentioned in our discussion of NO_2. In Panama, Lodge and Pate (131) were not satisfied with their analytical techniques, but they did state that NH_3 concentrations in this area were generally higher than 20 ppb. Junge (3) reported about 7 ppb NH_3 in Florida in 1954 and 3 ppb from the East Coast of Hawaii. At 10,000 ft in Hawaii he found 1.4 ppb NH_3. Junge (147) also reports on longer term NH_3 data from the Swedish air chemistry network. Monthly average values over a year's time at 4 stations showed variations mostly between 1 and 10 ppb. There was some indication of a summertime maximum, with a 4 station average of about 7 ppb in July and August and 2.5 ppb in December.

In the NH_3 data it would be reasonable to find an increase in concentrations coinciding with an increase in biological activity, and this would be consistent with the fact that the highest NH_3 concentrations were reported by Lodge from Panama.

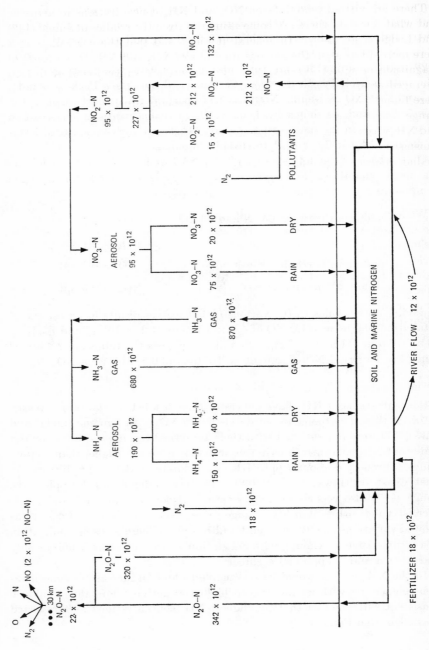

Fig. 13. Environmental nitrogen circulation in terms of nitrogen quantities present in nitrogen compounds (grams/year).

4. Ammonium and Nitrate Aerosols

There are virtually no data on NO_3 and NH_4 concentrations in aerosols, and what few data there are come primarily from the studies of Junge (147) and Lodge et al. (148).   In general, it was found that the particulate ions were more dilute than the gaseous fractions of NO_2 and NH_3 by an order of magnitude or more.   In his 1956 Florida studies Junge found that NO_3 averaged about 0.33 $\mu g/m^3$.   Further, the giant particles, $0.8\ \mu < r < 8.0\ \mu$ were found to contain almost all the NO_3 present.   In his Hawaii samples Junge found generally lower concentrations; average NO_3 was 0.08 $\mu g/m^3$ and NH_4 was 0.04 $\mu g/m^3$.   Differences by particle size were much less marked. During the sampling in the mid-Pacific, Lodge et al. (148) found an NO_3 median concentration of about 0.015 $\mu g/m^3$.

E. The Environmental Nitrogen Cycle

We have analyzed the cycle of nitrogen and nitrogen compounds through the atmospheric environment.   The estimate of the annual nitrogen cycle is shown in Figure 13 in terms of total nitrogen in the various fractions of the cycle.   The cycle expressed in terms of nitrogen component tonnages is shown in Figure 14.   The cycle for nitrogen is made up essentially of three separate cycles (the N_2O cycle, the NH_3 cycle, and the NO_x cycle).   There is a major difference between the revised natural NO emission cycle and our previous one (151) in that we have not found it necessary to include an association between NO_x and NH_3 phases of the cycle.

F. The Nitrous Oxide Cycle

The understanding of N_2O in our environment has improved in recent years.   Now the measurements of the ambient concentration seem to be fairly consistent and indicate a level of about 0.25 ppm (6,142).   Measurements have also shown that there is N_2O present in the ocean and that some mechanism in the ocean is providing a minor sink (125).   These facts are supplemented by newer calculations of the annual stratospheric photodissociation rate indicating a much lower value, 24×10^{12} g N_2O–N (142) compared with the previous rate estimate of about 6.3×10^{14} g/year made by Goody and Walshaw (123).

This previous estimate of N_2O destruction was generally balanced by production in the soil at a similar rate, 10^{15} g/year as N_2O, based on the soil production studies of Arnold (124).   Reactions in the soil have been summarized in detail by Bates and Hays (142), who point out the various factors, such as moisture and pH, that affect Arnold's N_2O production estimate.   They also note, however, that N_2O is destroyed by some microbiological reactions, i.e., those described by Wijler and Delwiche (152),

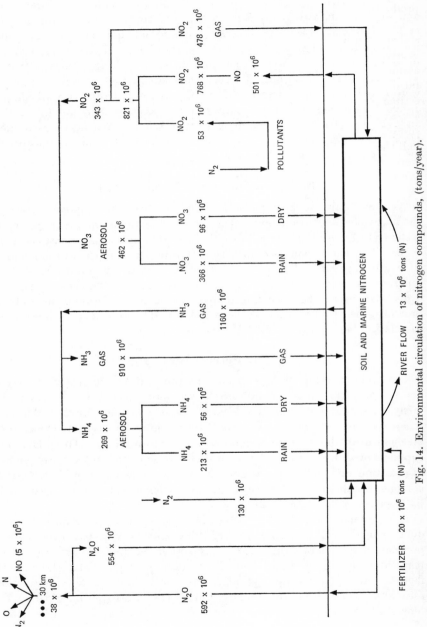

Fig. 14. Environmental circulation of nitrogen compounds, (tons/year).

with the resultant production of N_2. Burris (121) has reported that the presence of N_2O will inhibit the fixation of N_2. Thus, for N_2O there is both production and destruction at the soil surface.

In our estimate of the circulation of N_2O, both the stratospheric and the soil sinks have been considered. We also suggest the possibility that N_2O is absorbed in plant tissues during photosynthesis. For example, if we assumed absorption by vegetation in the same proportion as the atmospheric ratio with CO_2; in the case of N_2O this ratio is 300/0.25 or 1200 to 1, and then estimate the annual production rate of CO_2 as 6×10^{17} g on the basis of a 4 year CO_2 residence time (151). At this rate the proportional rate of consumption of N_2O would be 5×10^{14} g/year (500×10^6 tons/year) or about half the N_2O emission rate calculated by Arnold (124). However, his results were very approximate and a factor of 2 is probably not too large a difference. We thus have the following estimate for N_2O in terms of N_2O–N.

Total N_2O–N in atmosphere 1.3×10^{15} g

Total N_2O–N in oceans 0.25×10^{15} g

(Craig and Gordon, 1963)

Estimated photodissociation 2.2×10^{13} g

(Bates and Hays, 1967)

Estimated biosphere loss 3.2×10^{14} g

Total production (to balance) 3.4×10^{14} g

$$\text{Residence time} = \frac{1.55 \times 10^{15}}{3.4 \times 10^{14}} = 4.5 \text{ years}$$

This conclusion as to the residence time is about the same as that estimated by Junge (3). If we used an annual CO_2 cycle of 10^{18} g, as estimated from Rabinowich (103), we would have to postulate a greater N_2O loss to the biosphere and a shorter, 2.9 years, residence time for N_2O.

Basically, we suggest that the N_2O cycle is a balanced system of biological production and a biological sink with an additional stratospheric sink that accounts for less than 10 % of the total loss of N_2O.

G. Nitrogen Fixations

Nitrogen is estimated as entering the soil at a rate of 118×10^{12} g/year. This is a bacterial fixation process and the estimated rate is based on calculations by Hutchinson (153). This portion of the cycle does not relate to other portions except that soil nitrogen is a source of NH_3, NO_x, and N_2O.

H. The Ammonia Cycle

The ammonia and ammonium aerosols that are present in the atmosphere are due primarily to ammonia production in the soil. The sea surface has not been clearly identified either as a source or a sink; however, data of Lodge and Pate (131) may be interpreted as indicating an NH_3 source in the Caribbean area. The magnitude of the biospheric NH_3 source has not been measured; however, following our previous techniques we can estimate the probable emission from concentration data. If, as previously estimated, 6 ppb is a reasonable average for NH_3 in the ambient atmosphere, the total amount of NH_3–N in the atmosphere at any one time would be about 17×10^{12} g. Because of its high solubility, and its origin at the surface, NH_3 can probably be assumed to have a rather rapid turnover, perhaps about 1 week, or even less, similar to that of SO_2. Georgii and Weber (154) indicate on the basis of chemical sampling and rain analysis in Frankfurt, Germany, that NH_3 concentrations in the atmosphere were reduced by about 50 % by rain, while the reduction in SO_2 concentrations was 36 %, and NO_2 was reduced 24 %. Thus, the assumption of similar residence times for NH_2 and SO_2 is probably not unreasonable. If the residence time for NH_3 were about 1 week, this would indicate an annual production of about 50 times the atmospheric mass or 850×10^{12} g.

In Figure 13 the production of NH_3–N is set at 870×10^{12} g in order to balance the calculated scavenging mechanisms, namely, gaseous deposition and aerosol formation processes.

Gaseous deposition for NH_3 is calculated from a set of fairly simple assumptions (a deposition velocity of 1 cm/sec and a global average concentration of 6 ppb). The deposition velocity assumes that NH_3 is as reactive with vegetation, land, and water surfaces as is SO_2. This seems to be generally reasonable considering comparative data on NH_3 and SO_2 in rainwater, reported by Georgii and Weber (154). The average NH_3 concentration is an approximate median of scattered remote measurements. If there were more data to draw on it would be desirable to try to differentiate between land and ocean areas, because indications are that land areas are the major source of natural NH_3 production.

Aerosol deposition of ammonium compounds represents the second scavenging mechanism for atmospheric NH_3. Ammonia reactions can occur readily in cloud and fog droplets, and reactions between NH_3 and SO_2 in solution to form $(NH_4)_2SO_4$ are estimated to be a major scavenging process for SO_2. The NH_4–N mass that is estimated as being included in this phase of the cycle is 190×10^{12} g NH_4–N. This amount is 125 % of the estimated amount of NH_4–N present in precipitation (the 25 % is added to account for dry aerosol deposition). The amount of NH_4–N deposited on a global basis

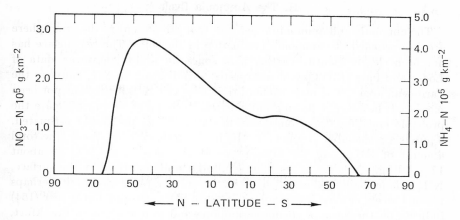

Fig. 15. NH_4-N and NO_3-N deposited by precipitation annually as a function of latitude. (Eriksson, 1952, Figs. 3 and 4).

in rain is estimated on the basis of Eriksson's (141) tabulation of NO_3–N and NH_4–N in precipitation.   These data indicate that there is a latitudinal distribution of NO_3–N and NH_4–N, with peak depositions occurring in the Northern hemisphere.   Figure 15 is our interpretation of the data presented by Eriksson (See Figures 4 and 5 and ref. (141)).   As indicated by Figure 15 the same latitude distribution is applied to both NH_4–N and NO_3–N.   The total deposition rate for NH_4–N was determined from Figure 15 by multiplying the deposition rate by the global area in $10°$ latitude belts.   The zero level of deposition in the polar regions is in general agreement with measurements of polar snow, especially in the Antarctic (155).   The total NH_4–N estimated to be brought to earth by precipitation is 150×10^{12} g/year.   This is a global rate and no differentiation is made between land and ocean areas.

I. The Cycle of Nitric Oxide and Nitrogen Dioxide

The third major phase of the nitrogen cycle is that entailing emissions of NO and NO_2 and the scavenging processes for both these gases and the nitrate aerosol resulting from reactions of NO_2 in the atmosphere.

The most readily recognized sources of atmospheric NO_x compounds are pollutant emissions.   We have estimated these emissions to be 15×10^{12} g/year NO_2–N.   Table 13 shows the distribution of NO_2 pollution sources.

As indicated by Figure 13 the major natural NO_x emission is estimated to be NO at a rate of 212×10^{12} g/year.   On the basis of our previously estimated average NO concentrations, this is equivalent to an NO residence time in the atmosphere of 4 days. This NO is scavenged readily by oxidation in the atmosphere, probably mainly through processes including ozone (129). Gaseous deposition of NO is considered to be negligible because of its low solubility and slow rate of reaction with surfaces. This natural NO_2 and the industrial NO_2 are scavenged both by gaseous deposition at a rate of 132 $\times$ 10^{12} g/year NO_2–N and through oxidation to nitrate aerosol at a rate of 95 $\times$ 10^{12} g/year NO_3–N. Gaseous deposition for NO_2 is based on our previously determined mean concentrations and a deposition velocity of 1 cm/sec. This deposition velocity is an approximation of experimental data obtained over alfalfa and oats by Tingey (156). Estimation of the precipitation of NO_3–N, 75 $\times$ 10^{12} g/year, is based on Figure 15 and Eriksson's (141) data, as was our estimate of NH_4–N deposition. Dry deposition of NO_3–N particulate material is estimated at 25% of the precipitation deposition.

J. Other Factors of the Nitrogen Cycle

The nitrogen cycle estimate is completed by our indication of the amount of fertilizer added to the soil, 18 $\times$ 10^{12} g/year as N, and the N transported from land to ocean by the rivers, 12 $\times$ 10^{12} g/year.

K. Discussion of the Nitrogen Cycle

This method for estimating the cycle for nitrogen does not require reactions of NH_3 to form nitrate to achieve a balance. This has been done mainly by relating the amount of deposition of NH_4–N and NO_3–N needed in the circulation model according to Eriksson's data (141) and Figure 15. Relatively low deposition rates in the Southern hemisphere of approximately 1 kg/10^4 m^2 for NO_3–N and NH_4–N are supported by data from Central Australia, reported by Wetselaar and Hutton (157). These changes in the circulation model seem to be reasonable and in line with available data. Since no satisfactory reactions are known for converting NH_3 to a nitrate compound, the fact that such a reaction is not needed in the calculations is a very significant development.

The recognition of a significant natural emission in the form of NO is also an important phase of this model. The total mass of the natural NO–N emission is about 80% larger than was estimated for natural NO_2–N emissions in our previous model (151).

The weak point in the nitrogen circulation model is primarily the small amount of available data that is applicable to the task. Differential data for land and ocean areas would be especially valuable in making more realistic estimates of the masses of material that enter into various portions of the cycle.

VI. HYDROCARBONS AND ORGANICS IN THE ATMOSPHERE

A. Introduction

Until the advent of serious air pollution in Los Angeles, California, in the late 1940's, air pollution was classically considered in terms of smoke and sulfur with perhaps some concern for CO where toxic hazards occurred.   The gradual gaining of understanding of the Los Angeles air pollution situation opened up new dimensions for air pollution problems.   These are characterized by photochemical reactions occurring in the atmosphere and involving organic materials and nitrogen oxides and producing as byproducts visibility-reducing aerosols, plant-damaging and eye-irritating organic compounds, and high concentrations of ozone and other oxidants.

Photochemical air pollution was once considered to be a unique attribute of Los Angeles or, at least, of California, and in many groups this affliction was considered a suitable retribution for the sins of "press-agentry" and "chamber-of-commercism."   However, in the past two decades photochemical air pollution has become an increasing occurrence in the world's major communities.   This spread of photochemical air pollution on a global scale was first recognized by Dr. Fritz Went in 1955 when he reported characteristic photochemical damage to plants in such diverse areas as Paris, France; Buenos Aires, Argentina; and New Delhi, India (158).

Thus, air pollution technology now of necessity must embrace both the inorganic contaminants of SO_2 and CO as well as a variety of organic compounds and the nitrogen oxides.   In this part of our discussion we will consider the organic phase of atmospheric chemistry and we will amplify the interactions between organic materials and the sulfur and nitrogen oxides which were presented in some of the preceding discussions.   As has been our pattern in the previous discussions, we will attempt to look at all of the organic materials which are emitted to our atmospheric environment, from both natural and man-made sources.

B. Sources and Concentrations of Organic Emanations

The sources of organic compounds in nature include emanations from the biosphere and from natural gas seepage.

Methane was first identified as a significant atmospheric constituent in infrared solar spectra by Migeotte in 1948.   These observations have since been confirmed by a number of others using both infrared spectra and chemical methods.   Concentrations of CH_4 in the atmosphere have generally fallen around 1.5 ppm in areas where pollution sources were not a factor. Junge (3) gives a summary of the studies which followed Migeotte's original idetinfication of CH_4 in the atmosphere.

More recent studies of CH_4 have included a limited number of vertical profiles by Bainbridge and Heidt (159). In these experiments, evacuated flasks were carried aloft by balloons and opened at predetermined altitudes to obtain the upper air samples. There were variations in the profiles but the CH_4 concentrations were nearly constant through the troposphere, decreasing at and above the tropopause.

In 1967 Stanford Research Institute carried out a limited period of atmospheric sampling of organic gases using gas chromatographic techniques at Point Barrow, Alaska (160). The results of one period of 24-hour sampling at Point Barrow are shown in Table XVII.

The component concentrations measured at the Point Barrow location should represent approximate atmospheric background levels. Certainly the CO and CH_4 levels are at the expected background levels. In addition to the fractional ppb concentrations of ethane, ethylene, butane, pentane, acetaldehyde, acetone, methanol, ethanol, benzene, and two unknown components, n-butanol was found at the surprisingly high concentration of 10 ppb. The presence of n-butanol at such high relative concentrations has not been confirmed, but if it is a real background component, it must be emitted by the biosphere.

Because of the dearth of clean atmosphere analyses for organic components other than CH_4, the Point Barrow data must serve as definitive at the present time.

With regard to CH_4, a number of different sampling studies do not differ significantly in their conclusions about average concentrations as compared to the earlier solar spectra data, and an average value for CH_4 of 1.5 ppm seems to be a good approximation for the ambient atmosphere.

The source of this CH_4 is primarily various bacterial decomposition reactions and it is produced in rather copious quantities in swamps, marshes, and other water bodies. The principal component of "marsh gas" is CH_4. Estimates of the magnitude of the CH_4 source are relatively difficult to make because of an almost complete lack of information. Koyama (161) estimates the production of CH_4 to be 310×10^6 tons/year from non-anthropogenic sources.

Natural gas has CH_4 as a major component and seepages are probable either in areas of considerable petroleum accumulation or in areas of oil field operations. However, analyses of the C_{14} content of atmospheric CH_4 has shown that it is composed of predominantly "young" carbon. This means that the biosphere rather than the old carbon from oil fields is the dominant source (162).

Urban areas are considerable sources for CH_4 as was shown by the analyses of Altshuller et al. (163) in Los Angeles and other cities. He measured CH_4 concentrations of 1.0 to 6 ppm and concluded that the elevated values

TABLE XVII

Concentrations of Atmospheric Contaminants at Port Barrow, Alaska

Date	Time								Concentration of atmospheric contaminant							
		Ethane ppb	Ethylene ppb	Butane ppb	Pentane ppb	Hexane ppb	Acetaldehyde ppb	Acetone ppb	Methanol-ethanol ppb	Benzene ppb	Unknown ppb	Unknown ppb	n-Butanol ppb	Methane ppm	Carbon monoxide ppb	Condensation nuclei N cc
9/2/67	2230	—a	0.6	0.6	0.6	—a	0.1	0.7	0.1	—a	—a	0.3	9.0	1.3	119	—a
9/2/67	2230	0.6	0.4	—a	0.6	—a	0.3	0.7	0.2	—a	0.3	0.3	9.0	NDc	100	
9/3/67	0030	0.4	0.3	—a	1.1	—a	0.3	1.0	0.2	0.1	0.1	0.3	10.0	ND	107	
9/3/67	0130	0.3	0.3	—a	1.3	—a	0.1	0.7	0.3	—a	—a	0.3	5.3	ND	80	
9/3/67	0230	0.4b	0.4b	—a	1.4	—a	0.3	0.8	0.3	0.4	—a	0.3	4.3	ND	119	200
9/3/67	0330	0.4	0.1	—a	0.4	—a	0.1	0.6	0.1	0.1	—a	0.3	9.0	1.48	123	100
9/3/67	0430	0.6b	0.6b	0.3	0.3	0.8	0.1	0.6	—a	—a	—a	0.3	10.0	1.50	90	
9/3/67	0530	0.3	0.4	—a	1.5	—a	0.4	0.8	0.1	0.1	—a	0.4	12.0	1.50	97	
9/3/67	0630	0.3	0.3	—a	2.2	—a	0.7	0.7	0.1	>0.1	>0.1	0.3	10.1	ND	110	
9/3/67	0730	0.6	0.3	0.2	0.4	—a	1.1	0.8	0.1	—a	—a	0.3	11.0	1.55	105	
9/3/67	0830	0.6	0.4	0.8	—a	—a	1.1	0.8	0.1	—a	—a	0.3	10.6	1.55	105	—a
9/3/67	0930	0.4	0.4	0.8	—a	—a	1.1	0.8	0.1	—a	—a	0.3	10.3	1.65	92	—a

a Not detected.
b The two compounds could not be separated.
c ND—No data, instrument difficulties.

TABLE XVIII

Estimated Release of Terpene-Type Hydrocarbons
from Vegetation[a]

Type of vegetation	Estimated emanations, tons/year
Coniferous forest	50×10^6
Hardwood forest	
Cultivated land	50×10^6
Steppes	
Carotene decomposition of organic material	70×10^6
	170×10^6

[a] Went 1960 (109).

resulted from natural gas leakage in distribution systems and from CH_4 found in combustion gases.

The biosphere is also a major contributor to the atmosphere of heavier hydrocarbons of the terpene class. Went (109) has made a careful study of the natural emanations of vegetation. Table XVIII is Went's estimate of the amount of terpene-type hydrocarbons that are released annually to the atmosphere. More detailed analyses of the heavier hydrocarbons in the ambient atmosphere have been reported by Rasmussen and Went (164). It is Went's theory that these terpene materials are polymerized in atmospheric photochemical reactions and form an organic aerosol and he attributes the "blue haze" found in many forested areas to the optical effects of this haze. Junge (3) estimates that these vegetation emanations can provide background concentrations of organic aerosol material in the range of 3 to 6 $\mu g/m^3$. We are not aware of any measurements of the ambient atmospheric aerosol which would provide a reliable check of this estimate.

Urban pollution also makes a contribution to the organics released to the atmosphere. Table XIX presents our estimate of the global emissions of organic pollutants. The total of almost 90×10^6 tons represents to a major extent petroleum processing and use with gasoline usage equalling 34×10^6 tons; refinery operations, 6.3×10^6 tons; petroleum evaporation and transfer losses, 7.8×10^6 tons; and solvent usage, 10×10^6 tons; for a total of 58×10^6 tons or 66% of the total. The estimate of solvent usage is a very rough one based on an estimate that world usage would be 3 times that in the United States. Incineration is also apparently a major contributor to organic emissions. We have used the same source estimate here as in our previous discussion. This estimate is roughly that the world total is estimated to be 5 times that of the United States, and the United States figure is estimated to be about 3 lb/person/day.   The fact that incineration has a poor combustion efficiency and thus a high calculated emission ratio of 100 lb/ton is a major reason that this source is so high.   If incineration were figured to be as efficient as the combustion of wood fuel, the estimated source contribution as shown in Table XIX would decrease by a factor of 30.

This estimate of 88×10^6 tons for pollutant hydrocarbons compares with the previously estimated CH_4 from natural sources of 310×10^6 tons and 170×10^6 tons of terpenes from vegetation.   Thus, total atmospheric hydrocarbon emissions are 568×10^6 tons, of which about 15% are due to pollution sources.

In areas where photochemical air pollution is a serious problem the major

TABLE XIX
Worldwide Hydrocarbon Emission Estimate

Source	Source quantity, tons ($\times 10^6$)	Emission factor, lb/ton	Percent reactive, %	Total emission, tons ($\times 10^6$)	Reactive emission, tons ($\times 10^4$)
Coal					
Power	1219	0.2[a]	15[b]	0.2	3
Industrial	1369	1.0[a]	15[b]	0.7	10.5
Domestic and commercial	404	10[a]	15[b]	2.0	30
Petroleum					
Refineries	11,317 bbl	56 tons 10,000 bbl[c]	14[b]	6.3	88
Gasoline	379	180[a]	44[b]	34	1500
Kerosene	100	0.6[a]	18[b]	<0.1	1
Fuel oil	287	1.0[a]	18[b]	0.1	1.8
Residual oil	507	0.9[a]	18[b]	0.2	3.6
Evaporation and transfer loss	379	41[b]	20[b]	7.8	156
Other					
Solvent Use[d]	3[d]	30 lb yr person[d]	15[b]	10	150
Incinerators[e]	500	100[a]	30[b]	25	750
Wood fuel	466	3[f]	15[f]	0.7	10.5
Forest fires	324	7[f]	21[f]	1.2	25
				88.3	2729.4

[a] Mayer 1965 (92).

[b] BAAPCD 1967 (165).

[c] Wohlers 1957 (139).

[d] L.A. data: 10^6 lb/day/6×10^6 people; U.S. total is about one half L.A. average, world total is 3 times U.S. total.

[e] Total is 5 times U.S. estimate.

[f] Darley et al. 1966 (98).

concern is with the olefins and other more reactive materials rather than just with the total organic emission. Using factors mainly derived by the Bay Area Air Pollution Control District (165), the total emissions in Table XIX have been broken down to give an estimate of the reactive hydrocarbon emissions. The result is that roughly one-third of the total is classed as reactive. Over half the reactive emissions are due to gasoline consumption.

C. Destruction Mechanisms and Sinks for Organic Compounds in the Atmosphere

1. Reactive Organics

The olefins from combustion sources and the terpenes from the biosphere undergo similar rapid chemical changes in the presence of nitrogen oxides and sunlight. The mechanisms and kinetics of olefin reactions in polluted atmospheres have been studied extensively. In a series of free radical polymeric reactions, reactive organics are transformed into condensed aerosol particles and eventually removed from the atmosphere by rainfall or surface deposition. The reaction series is initiated by the photolysis of NO_2, which produces atomic oxygen. In the absence of reactive hydrocarbons few secondary reactions follow, and the NO_2 photolysis reaction is not important. However, when olefins or other reactive organics are present in ppm concentrations, O atom attack on these molecules produces organo-oxy radicals, RO; these radicals in turn react with O_2 or nitrogen oxides, producing peroxy radicals and organo nitrates. The peroxy radicals react further to produce ozone. As the ozone concentration builds up, ozone oxidation reaction also becomes important both in the oxidation of NO to NO_2 and in the production of strong oxidants called peroxyacyl nitrates. The most important of these is peroxyacetyl nitrate (PAN). This compound causes plant damage at very low concentrations and probably is an important factor in the eye irritation potential of smog. The ultimate production of particulate material from these intermediates is strongly influenced by the presence of SO_2, although terpene-nitrogen oxide reactions appear to produce aerosol directly.

The lifetime of these organics as atmospheric gases is on the order of hours and the lifetime of the aerosols formed is only a few days.

The photochemistry of smog has recently been reviewed in detail by Altshuller and Bufalini (166) and Haagen-Smit and Wayne (144). The atmospheric terpene reactions are very similar.

2. Non-reactive Organics

Methane is the only important compound in this category, as the concentrations of the other aliphatics are very small relative to the concentration of methane.

While there is a great deal of work to be done before atmospheric methane sinks are understood and evaluated, there has been enough recent work, both direct and indirect, to make some assumptions about the fate of atmospheric methane.   Bainbridge and Heidt (159) estimated that only 10% of the atmospheric methane is transported into the stratosphere.   By analogy with atmospheric ammonia which is oxidized rapidly on the surfaces of vegetation, a similar large sink of methane is most probable.   However, the details and the effective oxidative agent have yet to be defined and identified.

D. The Atmospheric CH_4 Cycle

We have previously estimated the atmospheric residence time for various materials on the basis of atmospheric concentrations and total emissions. The only hydrocarbon for which there is sufficient information for this estimate is CH_4.

For CH_4 an average concentration of 1.5 ppm indicates a total mass of 4800×10^6 tons of CH_4 in the atmosphere.   Koyama's estimate of CH_4 emission is 310×10^6 tons (161) and provides an estimated residence time of about 16 years.   Ehhalt (162) has concluded that Koyama's CH_4 emission rate is actually a lower limit in that considerable areas of natural source areas were not included.   Koyama obtained about two-thirds of his CH_4 emission from biological action in paddy fields.   For these he had an estimated area of about 0.35×10^6 mi². Other obvious sources of natural CH_4 are the swamp lands of the world estimated to be about 10^6 mi² (167) and the hot humid tropical areas of the world which total about 11.4×10^6 mi² (168).   Thus, these other CH_4 source areas are some 35 times larger in area than Koyama's major source area.

It would seem that the CH_4 emission rate from swamp areas would probably be equivalent to that from paddy fields which, from Koyama's data, is 5.4×10^8 g/mi².   For hot, humid tropical areas it would seem that a CH_4 production rate of at least 10% of the paddy rate would not be out of line. From these considerations we have the following estimate of CH_4 production.

Paddy fields (Koyama)	$= 1.9 \times 10^{14}$ g
Other sources, mines, etc. (Koyama)	$= 0.8 \times 10^{14}$
Swamps	$= 5.7 \times 10^{14}$
Humid tropical areas	$= 6.1 \times 10^{14}$
	$\overline{14.5 \times 10^{14}}$ g
	or 1600×10^6 tons

If this emission rate is considered along with the 4800×10^6 tons of CH_4 in the atmosphere, the calculated CH_4 residence time becomes 4 years. Pollution sources, even if all of the estimated 61×10^6 tons of non-reactives were CH_4, would be a small factor in total atmospheric CH_4 concentrations.

The fact that this revised CH_4 residence time is now generally in the same range as has been estimated previously for CO, N_2O, and CO_2 is not, in our opinion, entirely accidental because we believe that the sink mechanisms for these less reactive materials are generally related. As we have mentioned previously, it is our opinion that these sink mechanisms are active in the biosphere both as bacterial reactions and as chemical reactions within plant tissues where CH_4 in respired air can be acted upon. A number of bacteria have been identified as methane producers (103). However, no scavenging processes have yet been identified with other vegetative cycles.

We have previously attempted to estimate the source production rates of less reactive compounds by assuming that the sources and sinks were in balance and that the sink reaction rate was proportional to the CO_2 reaction rate (151). The proportional factor for CO_2 and CH_4 is 300 ppm to 1.5 ppm or 200:1. Applying this factor to the previously used annual CO_2 sink of 10^{18} g (1.1×10^{12} tons) gives an estimated CH_4 sink of 5.5×10^9 tons or about 18 times Koyama's estimate and 115% of the CH_4 in the atmosphere.

It is our opinion that the CH_4 residence time is much shorter than the 16 years estimated by Koyama but probably not as short as the less than 1 year calculated from this CO_2 ratio. In support of this opinion it should be pointed out that the CH_4 background concentration measurements listed by Ehhalt (162) and by Bainbridge and Heidt (159) show considerable air mass and seasonal variation which would be indicative of a shorter residence time than Koyama's. In addition, Bainbridge and Heidt on the basis of very limited CH_4 data in the stratosphere estimated that the stratospheric CH_4 sink might account for only 10% of Koyama's production estimate.

The atmospheric CH_4 cycle obviously is an area where much more work needs to be done. However, it is our conclusion that the residence time is 1 to 4 years for CH_4 and that reactions in the biosphere are the major sink mechanisms. This would imply a source of about 2000×10^6 tons annually which does not seem out of line considering the proportion of swampy and tropical areas and the known paddy source areas.

VII. SUMMARY AND CONCLUSIONS

This analysis of the environmental cycles of gaseous atmospheric pollutants has considered three families of compounds: sulfurous, nitrogenous, and organics. In addition it has considered two inorganic carbon compounds: carbon monoxide and carbon dioxide. We have followed similar patterns in our analyses of these materials and have produced rather detailed analyses.

This analysis has included estimates of annual worldwide emissions of pollutants: SO_2, H_2S, CO, CO_2, NO_2, NH_3, and organics. We have also considered the magnitudes of the natural emanations of a variety of materials although it must be admitted that the means for estimating these emissions are very crude because so little study has been made of emissions from other than urban air pollution sources.

Important factors resulting from our analysis of atmospheric trace components and pollutants are summarized in Table XX. This table indicates the nature and size of the natural and pollutant sources, the estimated concentrations in the ambient atmosphere, probable scavenging reactions, estimated residence times, and the most important references which pertain to these various materials. This tabulation is, of course, amplified in the main body of this report.

In our discussions of atmospheric residence times for the various materials we have considered, it seems that the residence times seem to be either a few days or a few years. In the "few days" classification are the more reactive gases: SO_2, H_2S, NO, NO_2, and NH_3. In the longer "few years" classification are the significantly less reactive materials: CO, N_2O, CH_4, and CO_2. In the case of the more reactive gases the tropospheric scavenging and reaction processes are relatively well known and include photochemical, homogeneous, and heterogeneous reactions. For the latter materials, our low reactivity class, we postulate that the controlling scavenging process is in the biosphere and that the absorption and reactions for CO, CH_4, and N_2O are generally parallel. Although this is only a speculation at this time, our available facts in general support such an argument. For example, the calculated residence time for CO requires a tropospheric scavenging mechanism, and the biosphere is apparently the most logical one available. A similar argument holds for CH_4 although the magnitude of the natural source is not as well documented. For N_2O, a relatively short residence time of about 4 years would reconcile estimates of the likely rate of emission from the soil and the much slower photochemical dissociation of N_2O in the stratosphere.

The atmosphere of our environment obviously holds many secrets, including many things that we would like to know about the trace gases in the troposphere. It will require a considerable amount of research to expand our knowledge in this area because we will have to deal with large areas of the world and relatively small concentrations of material. However, the study of the unpolluted, ambient atmosphere must be an integral part of our program of air pollution research. Otherwise, we will be looking at only a very small part of our total environment.

TABLE XX

Summary of Sources, Concentrations, and Major Reactions of Atmospheric Trace Gases

Contaminant	Major pollution sources	Natural sources	Estimated emissions		Atmospheric background concentrations	Calculated atmospheric residence time	Removal reactions and sinks	Principal references	Remarks
			Pollution	Natural					
SO_2	Combustion of coal and oil	Volcanoes	146×10^6 tons	No estimate	0.2 ppb	4 days	Oxidation to sulfate by ozone or, after absorption, by solid and liquid aerosols	Eriksson (2, 53) Junge (3) Georgii (40)	Photochemical oxidation with NO_2 and HC may be the process needed to give rapid transformation of $SO_2 \rightarrow SO_4$
H_2S	Chemical processes, sewage treatment	Volcanoes, biological action in swamp areas	3×10^6 tons	100×10^6 tons	0.2 ppb	2 days	Oxidation to SO_2	Junge (3) Smith et al. (47)	Only one set of background concentrations available
CO	Auto exhaust and other combustion	Forest fires Oceans Terpene Reactions	304×10^6 tons	33×10^6 tons	0.1 ppm	<3 years	Probably soil organisms	Bates and Witherspoon (100) Swinnerton (87, 105) Robinson, Robbins (82, 83)	Ocean contributions to natural source probably low

Species	Man-made source	Natural source	Man-made amount	Natural amount	Concentration	Residence time	Removal mechanism	References	Remarks
NO/NO_2	Combustion	Bacterial action in soil (?)	53×10^6 tons	NO 430×10^6 tons, NO_2 658×10^6 tons	NO: 0.2–2 ppb, NO_2: 0.5–4 ppb	5 days	Oxidation to nitrate after sorption by solid and liquid aerosols, hydrocarbon photochemical reactions	Junge (3), Leighton (110), Lodge and Pate (131), Ripperton et al. (129), Hamilton et al. (130)	Very little work done on natural processes
NH_3	Waste treatment	Biological decay	4×10^6 tons	1160×10^6 tons	6 ppb to 20 ppb	7 days	Reaction with SO_2 to form $(NH_4)_2SO_4$ oxidation to nitrate	Eriksson (141), Georgii (27), Junge (3), Lodge and Pate (131)	Formation of ammonium salts is major NH_3 sink
N_2O	None	Biological action in soil	None	590×10^6 tons	0.25 ppm	4 years	Photo dissociation in stratosphere, biological action in soil	Bates and Hays (142)	No information on proposed absorption of N_2O by vegetation
Hydrocarbons	Combustion exhaust, chemical processes	Biological processes	88×10^6 tons	CH_4: 1.6×10^9 tons; Terpenes: 200×10^6 tons	CH_4: 1.5 ppm, non CH_4: <1ppb	4 years (CH_4)	Photochemical reaction with NO/NO_2, O_3; large sink necessary for CH_4	Went (109), Bates and Witherspoon (100), Ehhalt (162), Koyama (161), Cavanagh et al. (160)	"Reactive" hydrocarbon emissions from pollution = 27×10^6 tons
CO_2	Combustion	Biological decay, release from oceans	1.4×10^{10}	10^{12} tons	320 ppm	2–4 years	Biological adsorption and photosynthesis, absorption in oceans	Revelle (57), Keeling (63), Bolin and Eriksson (73)	Atmospheric concentrations increasing by 0.7 ppm/year

VII. ACKNOWLEDGMENTS

The preparation of the two reports upon which this review is based (169,170) was supported by the American Petroleum Institute and this is gratefully acknowledged.

Special commendation should also be given to our two librarians, Miss Elizabeth Feinler and Mrs. Caroline Arnold, for their detailed study of abstract sources, and to Dr. Christian E. Junge, Director, Meteorologisch-Geophysikalisches Institut, Johannes Gutenberg-Universität, Mainz, Germany, who acted as consultant to the American Petroleum Institute project.

References

1. U.S. Public Health Service, National Center for Air Pollution Control, *Air Quality Criteria for Sulfur Oxides*, USPHS Pub. No. 1619, Washington, D.C., March 1967.
2. Eriksson, E., Part I, *Tellus* **11**, 375–403 (1959); Part II, *Tellus* **12**, 63–109 (1960).
3. Junge, C. E., *Air Chemistry and Radioactivity*, Academic Press, New York, 1963.
4. Jensen, M. L., and N. Nakai, *Science* **134**, 2102 (1961).
5. Östlund, G., *Tellus* **11**, 478 (1959).
6. Junge, C. E., private communication, 1968.
7. Östlund, H. G., and J. Alexander, *J. Geophys. Res.*, **68**, 3995 (1963).
8. Sheehy, J. P., J. J. Henderson, C. I. Harding, and A. L. Danis, U.S. Public Health Service Publication No. 999–AP–3, 1963.
9. Rohrman, F. A., and J. H. Ludwig, *Chem. Eng. Prog.*, **61**, 59 (1965).
10. Stanford Research Institute, Projects 6508 and 6676, preliminary data.
11. U.S. Bureau of Mines, *Mineral Trade Notes*, **64**(12), 32–33, 48–49 (1967).
12. Katz, M., in *Air Pollution Handbook*, P. L. Magill, F. R. Holden, and C. Ackley, Eds., McGraw-Hill Book Company, Inc., New York, 1958.
13. U.N. Statistical Papers, Series J, No. 11, *World Energy Supplies*, 1963–1966, United Nations, New York (1968). (Tables 2 and 9).
14. McMahon, A. D., *Copper—A Materials Survey*, U.S. Bureau of Mines I.C. 8225 (1965), Table 26.
15. Herfindahl, O. C., *Copper Costs and Prices: 1870–1957*, Johns Hopkins Press, 1959, Tables 2 and A-1.
16. U.S. Bureau of Mines, *Materials Survey: Lead* (1951).
17. United Nations International Lead and Zinc Study Group, *Facts Affecting Consumption*, Appendix I (1966).
18. U.S. Bureau of Mines, *Materials Survey: Zinc* (1951).
19. Stanford Research Institute, *Fuel Usage Statistics* (unpublished), 1969.
20. Gerhard, E. R., and H. F. Johnstone, *Ind. Eng. Chem.*, **47**, 972–976 (1955).
21. Renzetti, N. A., and G. J. Doyle, *Int. J. Air Poll.*, **2**, 327 (1960).
22. Junge, C. E., and T. Ryan, *Quart. J. Roy. Meteorol. Soc.*, **84**, 46–55 (1958).
23. Johnstone, H. F., and D. R. Coughanowr, *Ind. Eng. Chem.*, **50**, 1169–1172 (1958).
24. Gall, D., *Oxidation of Sulfur Dioxide in Aqueous Solution*, paper presented at the OECD Symposium on Physico-Chemical Transformation of Sulfur Compounds in the Atmosphere and the Formation of Acid Smogs, Mainz, June 8–9, 1967.
25. van den Heuvel, A. P., and B. J. Mason, *Quart. J. Roy. Meteorol. Soc.* **89**, 271–275 (1963).
26. Gartrell, F. E., F. W. Thomas, and S. B. Carpenter, *Am. Ind. Hyg. Assoc. J.* **24** 113 (1963).
27. Georgii, H.-W., *J. Geophys. Res.* **68**, 3963 (1963).

28. Beilke, S., and H.-W. Georgii, *Tellus* **20**, 3 (1968).
29. Endow, N., G. J. Doyle, and J. L. Jones, *J. Air. Poll. Cont. Assoc.* **13**, 141 (1963)
30. Schuck, E. A., and G. J. Doyle, *Air Pollution Foundation Report No.* 29, 1959.
31. Harkins, J., and S. W. Nicksic, *J. Air Poll. Cont. Assoc.* **15**, 218 (1965).
32. Doyle, G. J., unpublished work, 1967.
33. Katz, M., and G. A. Ledingham, National Research Council of Canada, in *Effect of Sulfur Dioxide on Vegetation*, NCR No. 815, Ottawa, 1939.
34. Chamberlain, A. C., *Int. J. Air Poll.* **3**, 63 (1960).
35. Thompson, J. F., *Ann. Rev. Plant Physiology* **18**, 59 (1967).
36. Robbins, R. C., Development of methods for selective removal of oxidants from smog, Final Report, Project S-3200, Stanford Research Institute, for Agricultural Air Research Association, University of California at Riverside, January 17, 1961.
37. Cadle, R. D., and M. Ledford, *Air and Water Poll. Int. J.* **10**, 25–30 (1966).
38. Junge, C. E., E. Robinson, and F. L. Ludwig, J. App. Meteorol. **8**, 340–347(1969).
39. Junge, C. E., and N. Abel, Modification of aerosol size distribution in the atmosphere and development of an ion counter of high sensitivity, Gutenberg Univ., Mainz, Germany, Final Technical Report, 1965, Contract No. DA–91–591–EUC–3483.   DDC No. AD 469376.
40. Georgii, H.-W., and D. Jost, *Pure and Appl. Geophys.* **59**, 216 (1964).
41. Georgii, H.-W., J. Geophys. Res. **75**, 2365 (1970).
42. Cadle, R. D., J. Pate, and J. P. Lodge, paper presented to Am. Chem. Soc., Chicago, September 1967.
43. Kühme, H., paper to be published in Veröffentlichungen der Meteorexpedition 1965.
44. Cadle, R. D., W. H. Fischer, E. R. Frank, and J. P. Lodge, Jr., *J. Atmospheric Sci.* **25**, 100 (1968).
45. Meetham, A. R., *Quart. J. Roy. Meteorol. Soc.* **76**, 359–371 (1950).
46. Junge, C. E., *J. Geophys. Res.* **65**, 227–237 (1960).
47. Smith, A. F., D. G. Jenkins, and D. E. Cunningworth, *J. Appl. Chem.* **11**, 317 (1961).
48. Junge, C. E., C. W. Chagnon, and J. E. Manson, *J. Meteorology* **18**, 81 (1961).
49. Junge, C. E., *J. Geophys. Res.* **68**, 3975–3976 (1963).
50. Friend, J. P., *Tellus* **18**, 465 (1966).
51. Martell, E. A., *Tellus* **18**, 486 (1966).
52. Gruner, P., *Handbuch der Geophysik*, F. Linke and F. Möller, Eds., Bornträger, Berlin, Vol. 8, 1958, pp. 432–526.
53. Eriksson, E., *Tellus* **15**, 4001 (1963).
54. de Bary, E., and C. E. Junge, *Tellus* **15**, 370 (1963).
55. Gunn, R., *J. Atmos. Sci.* **21**, 168 (1964).
56. McCormick, R. A., and J. H. Ludwig, *Science* **156**, 1358 (1967).
57. Revelle, R., et al., Atmospheric carbon dioxide, in *Restoring the Quality of Our Environment*, Report of the Environmental Pollution Panel, President's Science Advisory Committee, November 1965.
58. United Nations, International Conference on Peaceful Uses of Atomic Energy, Proceedings, U.N., 1956.
59. Takahashi, T., *J. Geophys. Res.* **66**, 477 (1961).
60. Callendar, G. S., *Tellus* **10**, 243 (1958).
61. Pales, J. C., and C. D. Keeling, *J. Geophys. Res.* **70**, 6053 (1965).
62. Bischoff, W., and B. Bolin, *Tellus* **18**, 155 (1966).
63. Keeling, C. D., *Tellus* **12**, 200 (1960).
64. Lieth, H., *J. Geophys. Res.* **68**, 3887 (1963).

65. Junge, C. E., and C. Czeplak, *Tellus* **20**, 423 (1968).

66. Lieth, H. in, *Geographisches Taschenbuch*, Franz Steiner Werlag GmbH, Wiesbaden, 1965, pp. 72–79.

67. Young, J. A., and A. W. Fairhall, *J. Geophys. Res.* **73**, 1185 (1968).

68. Revelle, R., and H. E. Suess, *Tellus* **9**, 18 (1957).

69. Kaplan, L. D., *Tellus* **12**, 204 (1960).

70. Plass, G. N., *Tellus* **12**, 204 (1960).

71. Callendar, G. S., *Weather* **4**, 310 (1949).

72. Slocum, G., *Mon. Weather Rev.* **83**, 225 (1955).

73. Bolin, B., and E. Eriksson in, *The Atmosphere and the Sea in Motion*, B. Bolin, Ed., Rockefeller Institute Press, New York, 1959, pp. 130–142.

74. Manabe, S., and R. T. Wetherald, *J. Atmos. Sci.* **24**, 241 (1967).

75. Möller, F., *J. Geophys. Res.* **68**, 3877 (1963).

76. Fletcher, J. O., *Ice Extent on the Southern Ocean and Its Relation to World Climate*, The Rand Corp. Memo, RM–5793–NSF, March 1969.

77. Migeotte, M. V., *Phys. Rev.* **75**, 1108 (1949).

78. Migeotte, M. V., and L. Neven, *Physica* **16**, 423 (1950).

79. Shaw, J. H., R. M. Chapman, J. N. Howard, and M. L. Oxholm, *J. Op. Soc.* **113**, 268 (1951).

80. Adel, A., *Phys. Rev.* **75**, 1766 (1949).

81. Robbins, R. C., K. M. Borg, and E. Robinson, *J. Air. Poll. Cont. Assoc.* **18**, 106 (1968).

82. Robinson, E., and R. C. Robbins, *Antarctic J. of the U.S.* **3**, 194–196 (1968).

83. Robinson, E., and R. C. Robbins, Annals New York Academy of Sciences **174**, 89–95 (1970).

84. Cavanagh, L. A., C. F. Schadt, and E. Robinson, *Environ. Sci. and Tech.* **3**, 251–257 (1969).

85. Robinson, E., and R. C. Robbins, *J. Geophys. Res.* **74**, 1968–1974 (1969).

86. Robinson, E., L. A. Cavanagh, F. L. Ludwig, and R. C. Robbins, paper presented to Am. Chem. Soc., Houston, Texas, February 1970.

87. Swinnerton, J. W., V. J. Linnenbom, and C. H. Cheek, *Environ. Sci. and Technol.* **3**, 836–838 (1969).

88. Seiler, W., and C. E. Junge, *J. Geophys. Res.* **75**(12), 2217–2226 (1970).

89. Junge, C. E., Private communication (1969).

90. Seiler, W., and C. E. Junge, *Tellus* **12**, 447–449 (1969).

91. Stanford Research Institute, Project 6508, preliminary data, 1968 (unpublished).

92. Mayer, M., *A Compilation of Air Pollution Emission Factors*, U.S. Public Health Service, Division of Air Pollution, Cincinnati, Ohio, 1965.

93. U.S. Statistical Abstracts, 1967.

94. Parker, A., Estimates of Air Pollution in Great Britain in 1966, *Clean Air Yearbook, 1968–69*, p. 28., Nat. Soc. for Clean Air, London.

95. *Clean Air Yearbook, 1962–63.*, Nat. Soc. for Clean Air, London.

96. Heller, A. N., and D. F. Walters, *J. Air Poll. Cont. Assoc.* **15**, 423 (1965).

97. United Nations, *World Forest Inventory*, Food and Agricultural Organization, 1963.

98. Darley, E. F., F. R. Burleson, E. H. Mateer, J. T. Middleton, and V. P. Osterli, *J. Air Poll. Cont. Assoc.* **16**, 685 (1966).

99. U.S. Congress, *National Emission Standards Study*, Report of the Secretary, Health, Education, and Welfare, Senate Document No. 91-63, March 1970.

100. Bates, D. R., and A. E. Witherspoon, *Mon. Not. Roy. Astron. Soc.* **112**, 101–124 (1952).

101. Rankama, K., and T. G. Sahama, *Geochemistry*, University of Chicago Press, 1949.
102. Wilks, S. S., *Science* **129**, 964–966 (1959).
103. Rabinowitch, E., *Photosynthesis and Related Processes*, Vol. I, Wiley-Interscience Publishers, Inc., New York, 1945.
104. McElroy, J. J., *Calif. Agri.* **14**(9), 3–4 (1960).
105. Swinnerton, J. W., V. J. Linnenbom, and C. H. Cheek, *Environ. Sci. Limnol. Oceanog.* **13**(1), 193–195 (1968).
106. Barham, E. G., *Science* **140**, 826–828 (1963).
107. Pickwell, G. V., *Gas and Bubble Production by Siphonophores*, NUWCTP8, September 1967.
108. Swinnerton, J. W., V. J. Linnenbom, and C. H. Cheek, *Science* **167**, 984–986 (1970).
109. Went, F. W., *Proc. Nat. Acad. Sci.* **46**, 212 (1960).
110. Leighton, P. A., *Photochemistry of Air Pollution*, Academic Press, New York, 1961.
111. Went, F. W., *Tellus* **18**(203), 549–556 (1966).
112. Mellor, J. W., *A Comprehensive Treatise on Inorganic and Theoretical Chemistry*, Longmans and Co., London, 1924.
113. Zatsiorski, M., V. Kondrateev, and S. Solnishkovn, *J. Phys. Chem.* (USSR) **14**, 1521 (1940).
114. Brown, F. B., and R. H. Crist, *J. Chem. Phys.* **16**, 361 (1947).
115. Greiner, N. R., *J. Chem. Phys.* **46**, 2795–2799 (1967).
116. Doyle, G. J., Stanford Research Institute, 1968 (unpublished).
117. Douglas, E., *J. Phys. Chem.* **71**, 1931 (1967).
118. Jaffe, L. S., *J. Air Poll. Cont. Assoc.* **18**, 534–540 (1968).
119. Levy, E. A., *The Biosphere as a Possible Sink for Carbon Monoxide Emitted to the Atmosphere*, Final Report, Contracts CAPA 46812–68, and CPA 22–69–43, Project 7888, Stanford Research Institute 1969.
120. Ducet, G., and A. J. Rosenberg, *Ann. Rev. Plant Physiology* **13**, 171 (1962).
121. Burris, R. H., *Ann. Rev. Plant Physiology* **17**, 155 (1966).
122. Carr, D. J., "Chemical Influences of the Environment," in *Encyclopedia of Physiology*, **16**, 773–775 (1961).
123. Goody, R. M., and C. D. Walshaw, *Quart. J. Roy. Meteorol. Soc.* **79**, 496 (1953).
124. Arnold, P. W., *J. Soil Sci.* **5**, 116 (1954).
125. Craig, H., and L. I. Gordon, *Geochimica et Cosmochimica Acta* **27**, 949 (1963).
126. Altshuller, A. P., *Tellus* **10**, 479 (1958).
127. Lowry, T., and L. M. Schuman, *J. Am. Med. Assoc.* **162**, 153 (1956).
128. Peterson, W. H. *Production of Toxic Gas* (NO_2) *from Silage*, presented at 130th National Meeting, Am. Chem. Soc., Atlantic City, New Jersey, 1956.
129. Ripperton, L. A., L. Kornreich, and J. J. B. Worth, *J. Air Poll. Cont. Assoc.* **20**(9), 589–592 (1970).
130. Hamilton, H. L., J. J. B. Worth, and L. A. Ripperton, *An Atmospheric Physics and Chemistry Study on Pike's Peak in Support of Pulmonary Edema Research*, Research Triangle Institute, U.S. Army Research Office, Contract No. DA–HC19–67–C–0029, May 1968.
131. Lodge, J. P., Jr., and J. B. Pate, *Science* **153**, 408 (1966).
132. Elkin, H. F., in *Air Pollution*, Vol. II, A. C. Stern, Ed., Academic Press, New York, 1962.
133. Gerstle, R. W., and D. A. Kemnitz, *J. Air Poll. Cont. Assoc.* **17**, 324 (1967).
134. Wohlers, H. C., and G. B. Bell, *Literature Review of Metropolitan Air Pollutant Concentrations*, Final Report, Stanford Research Institute, Contract DA 18–064–404–CML–123, Project 1816, 1956.
135. Uhl, W. C., *World Petroleum Ann. Rev.* **36**, 28 (1965).

136. Gorle, E. F., in *Bacterial Physiology*, C. H. Werkman and P. W. Wilson, Eds., Academic Press, New York, 1951.

137. Allison, F. E., *Advances in Agronomy* **7**, 213 (1955).

138. Tebbens, B. D., in *Air Pollution*, second edition, A. C. Stern, Ed., Academic Press, New York, 1968.

139. Wohlers, H. C., Community Air Pollution Sources, Stanford Research Institute, 1957 (unpublished).

140. Hoering, T., *Geochim. Cosmochim. Acta* **12**, 97 (1957).

141. Eriksson, E., *Tellus* **4**, 215 (1952).

142. Bates, D. R., and P. B. Hays, *Planetary Space Sci.* **15**, 189 (1967).

143. McHaney, L. R. J., "The Vapor Phase Reactions Between Nitrogen Oxides and Water," MS thesis in chemical engineering, University of Illinois, 1953.

144. Haagen-Smit, A. J., and L. G. Wayne, in *Air Pollution*, second edition, A. C. Stern, Ed., Academic Press, New York, 1968.

145. Adel, A., *Science* **113**, 624 (1951).

146. Goldberg, L., and E. A. Mueller, *J. Opt. Soc. Am.* **43**, 1033 (1953).

147. Junge, C. E., *Tellus* **8**, 127 (1956).

148. Lodge, J. P., Jr., A. J. MacDonald, and E. Vihman, *Tellus* **12**, 184 (1960).

149. O'Connor, T. C., *Atmospheric Condensation Nuclei and Trace Gases*, Final Report, Dept. Physics, University College, Galway, Ireland, Contract No. DA–91–591–EUC–2126, 1962.

150. Fischer, W. H., J. P. Lodge, Jr., A. F. Wartburg, and J. B. Pate, *Env. Sci. and Tech.* **2**, 464–466 (1968).

151. Robinson, E., and R. C. Robbins, Gaseous Atmospheric Pollutants from Urban and Natural Sources, in *Global Effects of Environmental Pollution*, S. F. Singer, Ed., D. Reidel Publishing Company, Dordrecht-Holland, 1970.

152. Wijler, J., and C. C. Delwiche, *Plant and Soil* **5**, 155 (1954).

153. Hutchinson, G. E., in *The Earth as a Planet*, G. P. Kuiper, Ed., University of Chicago Press, 1954.

154. Georgii, H.-W., and E. Weber, The chemical composition of individual rainfalls, Goethe Universität, Frankfort, Tech. Note. AFCRL–TN–60–827, Contract AF 61(052)–249, 1960.

155. Wilson, A. T., and D. A. House, *J. Geophys. Res.* **70**, 5515–5518 (1965).

156. Tingey, D. T., "Foliar Absorption of NO_2," MS thesis, Dept. of Botany, Univ. of Utah, Salt Lake City, Utah, June 1968.

157. Wetselaar, R., and J. T. Hutton, *Aust. J. Agri. Res.* **14**, 319–339 (1963).

158 Went, F. W., in Proc. 3rd National Air Pollution Symposium, Pasadena, Calif., Stanford Research Institute, April 1955.

159. Bainbridge, A. E., and L. E. Heidt, *Tellus* **18**, 221 (1966).

160. Cavanagh, L. A., SRI Project 6689, 1967 (unpublished).

161. Koyama, T., *J. Geophys. Res.* **68**, 3971 (1963).

162. Ehhalt, D. H., *J. Air Poll. Cont. Assoc.* **17**, 518 (1967).

163. Altshuller, A. P., G. C. Ortman, B. E. Saltsman, and R. E. Neligan, *J. Air. Poll. Cont. Assoc.* **16**, 87 (1966).

164. Rasmussen, R., and F. W. Went, *Science* **144**, 566 (1964).

165. Bay Area Air Pollution Control District, San Francisco, Calif., unpublished source tabulations, 1967.

166. Altshuller, A. P., and J. J. Bufalini, *Photochem. Photobiol.* **4**, 97 (1965).

167. Twenhofel, W. H., *Principles of Sedimentation*, McGraw-Hill Book Company, Inc., New York, 1950.

168. United Nations Educational, Scientific, and Cultural Organization, *Problems in Humid Tropics*, Paris, 1958.

169. Robinson, E., and R. C. Robbins, *Sources, Abundance, and Fate of Gaseous Atmospheric Pollutants*, Final Report Stanford Research Institute, Project PR–6755, for American Petroleum Institute, New York, 1968.

170. Robinson, E., and R. C. Robbins, *Sources, Abundance, and Fate of Gaseous Atmospheric Pollutants*, Supplemental Report Stanford Research Institute, Project PR–6755, for American Petroleum Institute, New York, 1968.

The Law and Air Pollution

ANNEMAREE LANTERI

*University of Melbourne
Victoria, Australia*

I. INTRODUCTION

Although it appears (1) that a form of legislative control of air pollution existed as early as the fourteenth century, the magnitude of the problem of conservation of air quality has become really apparent only since the industrial revolution.   Early regulation was limited to proscription of the production of smoke and offensive odors.   However, due to the development of technology and industrial processes, the sources of less obvious forms of pollution, as well as the amount of pollution itself have increased.   Today the field of air quality management is a complex of sociological, legal, and technological problems.   Legislation which aims at effective control must balance the social desirability of encouraging progress in a consumer-oriented society with the requirements for a healthy and aesthetically satisfying civilization.   It must achieve this balance while remaining within the bounds of legal feasibility and considering the development rate of technical skills and processes.   For these reasons any legislative scheme should be preceded by soundly based research covering many disciplines.   This chapter is primarily concerned with the general principles involved in the legal control of air pollution and a survey of current legislation in a number of countries.

II. PRIVATE REMEDIES

The earliest legal actions brought to abate air pollution were based on private law, and arose when a person who had suffered some loss because of circumstances created by another's activities sought compensation from him. The United Kingdom, the United States, Canada, Australia, and New Zealand are among the countries whose legal systems are based on common law.   The term "common law" refers to that body of law which is derived from previously decided cases and not founded upon legislation.   It is distinguishable from the Civil Law of France, Germany and other Continental countries, which is based on a legislative consolidation or codification. Several private actions are relevant to a discussion of the legal control of air pollution.

A. Common Law Actions

There are four main causes of action which are relevant; they are: actions for damages due to the negligence of the defendant, for damages due to the escape of a dangerous thing from the defendant's property, and actions for public and private nuisance.   These actions may be adopted for use in a variety of fact situations.   It should be noted that there will be variations in their application and development in the countries where they are known, although their basic characteristics remain the same. A brief description of these actions may help to clarify the role of the common law in this field.

To succeed in an action for damages arising out of the negligent emission of a pollutant, the plaintiff (the person seeking compensation) must first show that the defendant ought to have foreseen that the plaintiff could be affected by such emission. For instance, a factory operator would be expected to realize that his employees and neighbors in the immediate vicinity will probably be affected by his activities. The plaintiff must then show that the defendant has in fact acted negligently and, because of this negligence, the plaintiff has suffered damage to his person or property.

In an action for damages caused by the escape of a dangerous thing, the plaintiff must show that the defendant has brought on to his land something which is not naturally there and which is likely to be dangerous if it escapes. A variety of objects have been held to be within this category, for instance water, fire, gas, oil, explosives, poisonous trees, and a flagpole. If it has escaped and injured the plaintiff or his property, then he may recover damages whether or not this event was foreseeable by the defendant. This type of action is available to a plaintiff who cannot prove negligence in the conduct of the defendant but who can prove the emission of a toxic pollutant.

The action of private nuisance is available to a landholder whose enjoyment is diminished by the defendant's use of his own premises. The plaintiff must show that the defendant has measurably spoiled his use and quiet enjoyment of his property and that this has occurred more than once. The court may balance the conflicting interests of the parties in reaching its decision. If it is socially undesirable that the defendant be forced to pay vast sums of compensation, or to close down his operations entirely, then the plaintiff will fail. If the balance of interests is weighed in favor of the plaintiff, and he can show that the nuisance is very likely to recur, then he may be granted an injunction to prevent the defendant from continuing the offensive operations under penalty of contempt of court.

In certain circumstances the Attorney General or District Attorney may bring an action on behalf of the public at large. This type of nuisance must have affected a widespread area and more people than a neighboring landholder. In such a case, a private individual who has sustained injuries may sue, although he is not a landholder and has no proprietary interest which has been damaged. Apart from this situation, however, nuisance is limited to the landholder himself and does not even extend to members of his family or household.

As a general rule, an employer is liable to compensate for injuries caused by the intentional or negligent actions of his employees which take place in the course of their employment. Thus, an action may be brought against the person primarily at fault and if necessary, against his employer.

There may be other causes of action which could also be pressed into service. An action could be brought for trespass to land or air space where pollutants travelled across or descended onto the plaintiff's property. It has also been suggested that actions against a governmental unit for failing to take reasonable steps to control pollution within its jurisdiction may succeed (2). An action against the manufacturer of motor vehicles which emit pollutants may succeed as an extension of liability for the production of negligently designed or constructed goods. Given the flexibility of development of the common law and increased public awareness of the dangers involved in unregulated air pollution, it is beyond question that more private actions will be brought against polluters.

B. Civil Law Actions

The French Code, "the Code Civil," recognizes a fully generalized principle of liability for one's actions which either intentionally or negligently cause injury to another. Articles 1382 and 1383 of the Code Civil provide:

"Any act by which a person causes damage to another makes the person by whose fault the damage has occurred liable to make reparation for it. Everyone is liable for the damage he has caused not only by his acts but also by his negligence or imprudence." Thus, a defendant will incur liability if his conduct is such that fault can be imputed to him, and the damage which is recognized as the subject matter for compensation is any harm done to the plaintiff's person or property. Article 1384 [5] provides that employers are liable for damage caused by their employees in the course of their employment.

The German Code, or "Bürgerliches Gesetzbuch (B.G.B.)," provides in Article 823 that a person who intentionally or negligently injures unlawfully the life, body, health, freedom, property, or other rights of another is bound to compensate him for the damage caused. A person who infringes a statute intended for the protection of another incurs the same liability. However, under Article 831 of B.G.B., an employer is only liable for his employees' actions if he himself was at fault in the selection or supervision of them. The degree of care required to exonerate an employer varies with the danger of the work undertaken. It appears therefore, that a plaintiff in France or Germany could recover for damage caused by pollution in much the same circumstances as he could under the common law.

C. The Significance of Private Remedies

The significant characteristic of these remedies, which clearly indicates that alone they are hopelessly inadequate to preserve a satisfactory standard of air quality, is the fact that they are designed to compensate rather than to prevent. The injunction is only a limited exception to this proposition.

In the vast majority of cases, concentration is focussed on monetary reparation rather than prophylactic measures. Practically speaking, relatively few people in the community are aware of their rights and fewer still attempt to enforce them. Some of this disinclination to rely on legal action is justified due to the expense, problems of proof of causation, and temporal limitation of actions. On the other hand, the common law actions provide developed methods of compensation in a field where statutory provisions have been generally limited to prescribing and enforcing emission standards. They are not limited to industrial or commercial sources but are available against any offender. A streamlined procedural system could make the common law an effective additional weapon against air pollution, provided courts were willing to award substantial sums for damages. An intensive public education program would be necessary. It may be desirable to make it a defense at common law to show that prescribed standards have not been exceeded, and to make the facilities of local enforcement agencies available to prospective litigants who need accurate measurements of air quality and flow (3).

III. LEGISLATIVE CONTROL

The increase in the number of sources and rates of atmospheric pollution which resulted from the development and expansion of technology, highlighted the inadequacy of private litigation as a means of air quality control in the twentieth century. Growing public concern over apparent pollution, coupled with a series of disasters resulting in considerable loss of human life, put pressure on legislative bodies to turn their attention to the legal control of atmospheric pollution through specific legislative programs.

The earliest forms of legislation in this field were usually based in the power of the government to regulate and prohibit activities creating public nuisances. This type of legislation either prohibited the emission of smoke or odors which constituted a nuisance at common law, or declared a certain standard of emission to be a public nuisance *per se* (4). The inadequacy of this legislation to combat growing pollution centered around three major reasons. The basic flaw in its construction was that it attempted to control simply through punishment where and when pollution was discovered, rather than attempting to prevent the discharge of pollutants at their source. It only applied to limited sources of pollution and the consequent exemption of private chimneys and incinerators and other nonindustrial pollution did not provide a wide enough basis of attack to ensure efficient control. Since enforcement and administration of this type of legislative scheme was left largely in the hands of local authorities, its enforcing officers were generally untrained, uncoordinated and overworked.

More recent legislation has ignored the concept of nuisance and simply declared that the emission of pollutants above the maximum permissible limit was an offense. Basically the characteristics of modern legislative programs include: a central enforcement and appeal agency, which allows the appointment of trained officers of air pollution control; a list of premises which require permits to operate, the submission of plans for the repair, alteration or construction of new plants to the agency for approval; and, the periodic inspection of plants by enforcement officers. Apart from the list of scheduled premises, the regulatory powers of the enforcement agency often cover other nonindustrial or commercial sources of pollution, such as hospitals and public works as well as the use of fuel and the creation of protective zones. This type of approach aims primarily at prevention rather than punishment and is thus likely to be much more effective than the more primitive, punitive approach. The validity of this type of legislation depends solely upon the ambit of legislative power exercised by the body which enacts it, rather than judicial interpretation of the law of nuisance.

The power of a legislature to regulate behavior may be limited by a variety of considerations. The exercise of legislative powers by a body such as a city or county council will be limited by the extent of the power delegated to it by its immediate legislative superior (the state or national government). If a local council has had power delegated to it from the state government to make regulations for the control of "smoke dust and grit emissions from industrial sources," it could not validly make a regulation which covered the emission of invisible fumes or gases or emissions from domestic sources. Although a law or regulation is presumed valid until successfully challenged, as a general rule, money paid to governmental authorities under a regulation later found to be invalid is not recoverable (5).

In a federation of states the individual states may be limited in their power by the federal constitution, as well as by the terms of their own constitutions. For example, the Australian state of Victoria is limited by its Constitution to making laws which are "in and for" the state, in the sense that they must have some connection with the state. Furthermore, any Victorian law which is inconsistent with a federal law will be invalid to the extent of the inconsistency by virtue of Section 109 of the Commonwealth Constitution.

Two basic methods of legislative control have been used in attempts to prevent and abate air pollution. The first aims at control within the limits imposed by economic and technological factors. It follows the form of a broad law framed in fairly general terms, made by the central government either of the state or country, which sets out the basic scheme for control. Power to make regulations or decrees which specifically implement the policy of the basic law is delegated either to a centralized goverment body or perhaps to local or regional authorities. The regulations or decrees deal with technical matters such as the prescription of standards of emissions, the listing of premises required to comply with prescribed standards, the description of air cleaning devices which may be required to be used, and the delineation of smokeless zones or protective belts. Generally, they are made on the recommendations of a group or groups of experts. The regulatory body is of course limited in its powers by the terms of the basic enabling law. This type of legislative control program has the advantage that it avoids the need to change the basic law whenever a particular control measure is brought up to date with technical developments.

The second method is to embody any technical requirements and precise standards in the main law itself. If any delegation is made it is limited to responsibility for surveillance and enforcement. Although this is a less flexible approach, it may be equally efficient depending upon the administrative unit it is designed to serve and the standard of technical equipment and staff of the authorities required to carry out its enforcement.

A survey of national legislation reveals a variety of combinations of these two basic methods of legislative control, and the efficiency of any one of these combinations is of course concerned with the political structure and technical development of the unit it is to cover, as well as purely legislative considerations.

In the United States, community level action predates and usually surpasses in expenditure schemes introduced by the states themselves. Naturally, there is a wide variation in the methods of control chosen by local authorities because they are tailored for special local conditions. They range in development from the sophisticated scheme of Chicago, Ill. with an annual per capita budget of 32.7c to that of Memphis, Tenn. with a similar budget of 0.8c. Although the number of local and regional programs rose from 85 in 1961 to more than 130 in 1966, of the nearly 600 counties with populations over 50,000, less than 90 had control programs at that date. Only 3 of the 50 largest cities in the United States (Los Angeles, Cal., Long Beach, Cal. and Akron, Ohio) are spending more than 40c per capita (6).

Regional schemes based on the coordinated efforts of several local authorities appear to be a highly efficient method of control. The San Francisco Bay Area Air Pollution Control District is an example of this type of scheme aimed at controlling pollution within a geographical area rather than one based on municipal boundaries. In Australia local schemes are non-existent. Although local councils generally have power to regulate nuisance conditions, there is no instance of a specially adapted scheme to abate air pollution below the state level.

Legislative activity at state level in the United States is comparatively recent and thus undeveloped compared with local progress. In 1965, although 40 states had budgets of $5000 or more for air pollution control programs, only 9 of these states were actually involved in regulatory as opposed to information-gathering and advisory activities. California conducts the most expensive and complex control action at state level. Nevertheless, it may be that state control will assume a more dominant role in the future, supported by federal activity. In Australia the lead in clean air legislation has been taken by the states, although the development of control measures is embryonic compared with progress in the United States.

In federations such as the United States and Australia national legislation will take a different form and play a different role than in countries with completely centralized authority. In the United States federal involvement is increasing, although, apart from regulation of automotive pollution, its aim is to encourage and assist state and local programs. The most recent advance in federal legislation was the Air Quality Act of 1967, which *inter alia* provides for the future control of pollution through air quality control regions. This may prove to be the most rational approach to a national problem, although it will be some time before it is fully implemented and its success calculated.

IV. GENERAL SURVEY OF NATIONAL LEGISLATION

In the years following the end of World War II national legislative programs to abate and control air pollution have appeared in ever increasing numbers and embody progressively more sophisticated measures. This survey is of a general nature and covers those countries which have published legislation in this field. The *International Digest of Health Legislation* (*I.D.H.L.*) has published either summaries or reprints of most of this legislation, and the references below indicate both the legal citation and the relevant Digest report (6a).

When considering any national legislation of this nature it should be borne in mind that the enabling act or decree, which is usually framed in terms of a general policy statement, does not positively indicate the efficiency of air pollution control in that country. The effectiveness of these laws depends on several factors. Among these are the regulations passed to implement the law, the budget assigned to the department administering both law and regulations, the training and enthusiasm of the officers enforcing the provisions, and the amount of cooperation received from the public. The practical efficiency of a program may be revealed in air quality measurements taken over a reasonable period of time, although even these figures may reflect changes in economic and social habits not necessarily referable to the legislation under discussion. Nevertheless, legislation is the first step towards environmental conservation and therefore deserves careful consideration.

A. Australia

Power to legislate with respect to air pollution control rests at least primarily with the states. Although it is arguable that the Federal Government could enter this field of legislative activity it has not yet done so. However, a Senate Select Committee has recently tabled a report on air pollution in Australia and the recommendations made by it may result in the Federal Government assuming a leading role in future legislative development (6b).

Victoria passed a Clean Air Act in 1957, to be administered by the Commission of Public Health (7). The Act provides for the establishment of a Clean Air Committee charged with conducting investigations into the problems of air pollution and reporting its findings to the Minister of Public Health with suggestions for measures for control and abatement. Although the Act does not cover private dwellings, it does make provision for the regulation of the emission of dark smoke and air impurities, the height of chimneys and the approval of a new plant. Other regulatory powers of the Governor-in-Council under the Act relate to the authorization of fuel usage and the installation of air cleaning devices in industrial plants. The Commission may delegate its powers in respect of any of the provisions of the Act to local municipal councils, and this was done in 1958 in respect of the section dealing with the prohibition of dark smoke from industrial sources. Standards have been set for the emission of air impurities and are generally similar to those also adopted by regulation in New South Wales.

In December 1970 the Victorian Parliament passed the Environment Protection Act (No. 8056). This legislation provides for the control of all forms of pollution of the environment through a system of licensing of all sources of pollution. The Act is to be administered by the Environment Protection Authority with the assistance of an Environment Protection Council which comprises representatives of relevant fields of expertise and interest. Provision is made for the establishment of a State Environment Protection Policy to provide a frame of reference and basis for decisions of the Authority in its administration of the Act. When fully in force—only the general administrative provisions were proclaimed on March 15th, 1971—the Act should provide a comprehensive basis for effective and integrated environment protection activities.

The Clean Air Act of New South Wales (8) provides for a list of "scheduled premises" which are required to be licensed to operate. With regard to these premises the Act is to be implemented by the Department of Public Health, while others are dealt with by local councils, although the Department may take action where necessary. Provision is made for the approval of new plants either by the Department of the local authority, and an Air Pollution Advisory Committee is established, with the duty of general investigation of problems and advising the Under Secretary on desirable control measures.

Both Queensland and Western Australia have adopted legislation which is substantially the same as that of New South Wales. South Australia amended the Health Act of that State by inserting provisions of a general nature. They provide for the establishment of a committee to advise the Governor, and give him power to make regulations concerning the regulation and control of the emission of air impurities along lines similar to those of the Victorian Act.

B. Belgium

Until 1964, prohibitions against air pollution were scattered among several acts and local regulations. They were limited in scope and difficult to implement, and offenses generally incurred inadequate penalties. The law of December 1964 (9) empowers the Crown to take all measures appropriate for the control of atmospheric pollution whether arising from industrial, vehicular or domestic sources, and in particular to prohibit certain forms of pollution. Orders under the Act are to be proposed by the Minister of Public Health, and, depending on the sources of pollution, the Ministers for Labor, Public Works and Transport. The Minister for Public Health coordinates activities of authorities regarding the taking of samples, analysis and research into the effects of pollution, and public education. The law and orders thereunder are administered by appointed officers who have power to enter premises, collect samples, carry out tests and regulate the use of control devices. There is provision for emergency measures to be taken if any source threatens public health.

C. Brazil

The National Council for the Control of Environmental Pollution (hereinafter referred to as the N.C.C.E.P.) was established in 1967 (10). The decree which creates this body deals in general terms with problems of pollution of the environment and defines "pollution" broadly but with admirable comprehensiveness. Substances emitted by industrial, commercial or agricultural sources, or by machines and vehicles, must not be discharged into the atmosphere so as to cause pollution. Limits for these discharges may be specified either by the establishment of standards for emissions or by standards of air quality. The provisions of the decree are to apply to all sources of pollution whether under public or private law.

The N.C.C.E.P. is an organ of the Ministry of Health and is charged with the following duties *inter alia:*

1. to study, approve, and revise standards for the control of environmental pollution at both a national and regional level

2. to standardize working procedures for the control of such pollution

3. to lay down general standards for the control of pollution caused by industrial, commercial, and agricultural establishments, incinerators and internal combustion engines and vehicles fitted with such engines and other possible sources of pollution of the atmosphere

4. to encourage and support private investment in control plans

5. to organize national plans for the control of environmental pollution and set up a program for their implementation

6. to determine degrees of responsibility where more than one establishment is causing pollution

7. to arbitrate and act as an appeals body on interstate problems

8. to promote research

9. to organize public education regarding environmental pollution and to guide other agencies whether or not under State control.

D. Bulgaria

Under the general decree of October 1963 the Minister for Public Health and Social Welfare is charged with prime responsibility for air purity(11). Provision is made for the control of pollution by requiring authorization for plant design and construction so as to minimize air contamination. Purification devices must be installed in existing plants within a time specified by the Council of Ministers, and such devices must· be maintained and operated efficiently. The Council of Ministers may make exceptions to these principles in order to meet economic emergencies, but if it does so, then additional steps must be taken if necessary to protect health and cleanliness.

Regulations were passed in September 1964 to implement this general law under the direction of the Minister for Public Health and Social Welfare (12). Standards of emission of pollutants are to be approved by the State Sanitary Inspectorate of the Ministry of Public Health and Social Welfare, and the regulations deal with the methods for calculating such standards. Provision is made for wooded zones and green spaces to surround built up areas. Chimney heights are to be determined according to their pollution emissions, and much stress is placed on the appropriate selection of sites for industrial and other undertakings in built-up areas. Air cleaning equipment must be in service and properly maintained in any process likely to cause pollution, and additional measures for abatement may be ordered. Section

22 deals with motor vehicle management in large towns, providing for careful routing of heavy traffic, removal from service of defective vehicles, and
road planning to eliminate as much as possible the risk of vehicles remaining
idling unnecessarily.   Vehicles which give off visible fumes may be forbidden
access to large towns.   Responsibility for these provisions rests with the
Minister of the Interior, Transport, and Communications, the State Committee
of Construction and Architecture, the Minister for Public Health and Social
Welfare and the people's councils.

E. Canada

Legislative control of air pollution has been left almost entirely in the hands
of the Provinces.   The exceptions are the federal regulation of pollution
from vessels in navigable waters and from railways engaged in inter-
provisional transportation.   The former regulation is made under the
Canada Shipping Act 1952 and the latter under the Railway Act.   A Con-
vention between Canada and the United States regulates emissions from the
smelter at Trail, British Columbia.

Ontario passed an act for the control of atmospheric pollution in 1958 (13).
It delegates power to local municipalities to pass by-laws to control the emission from any source, of any type of air contaminant.   The by-laws must
first be submitted to the Minister for Health for approval.   The Minister
also has general powers to arrange research into the field of atmospheric
pollution.   Provision is made for the submission to municipalities for
approval of plans for the repair, installation and use of any equipment likely
to produce air contaminants.   A certificate that such equipment complies
with the relevant by-laws must be obtained before use.

The Province of Nova Scotia amended in 1960 the Municipal Act 1955 (14).
Local councils were given additional powers to make by-laws regulating the
control of emissions of smoke, odors and gases, to provide standards of emissions, to prohibit emissions in excess of such standards if a nuisance, to require
installation of control equipment, to authorize officers to implement these
requirements and to require owners, occupants, managers, and agents to
conduct tests and report thereon.

In 1960 Manitoba also amended the Public Health Act Regulations of
1945 (15).   The new Regulation defines "air contaminant" as being "dust,
fly ash, fumes, smoke or gases or any combination thereof."   These terms
are further defined, and the Regulation lays down maximum permissible
standards for the emission of various contaminants, as well as making it
unlawful to allow the discharge of any smoke, dust, cinders, fumes, gases or
offensive odors to the detriment or annoyance of others.

Alberta, in 1961 amended the Alberta Regulation No. 572/57 by inserting

Division 14 (16). This requires the submission for approval of plans for pipelines and certain specified industrial plants. Allowed densities of smoke and dust emissions are specified, and toxic or noxious materials may not be discharged without the written approval of the Provincial Board of Health.

Most recently, Saskatchewan, in 1965 passed an Air Pollution Control Act giving certain general powers to the Lieutenant Governor-in-Council (17). He may *inter alia* regulate the control the emissions of industrial sources of pollution, require the submission of plans to the provincial officer, divide the Province into districts and the districts into divisions for the prescription of maximum concentration of pollution standards. The Minister of Public Health may investigate air pollution problems, and municipalities may make by-laws prohibiting pollution from fuel-burning equipment and open fires. They may also require the submission of plans and designs for fuel-burning equipment and incinerators. An Air Pollution Advisory Committee is to be appointed by the Lieutenant Governor-in-Council to recommend to the Minister possible measures for abatement of air pollution.

Action has been taken at a municipal level in a number of Canadian cities. The regulations of May 1957 of the Municipality of Metropolitan Toronto may be taken as an example. They cover the emission of smoke, dust, flyash, soot, fumes, or other solid or gaseous products of combustion, and provide for the regulation of chimney heights, the issue of permits for fuel-burning equipment, the inspection of premises and equipment and the investigation of complaints and the study of smoke conditions. The town of Trafalgar (Ontario) has enacted regulations of a special character dealing specifically with the control of pollution caused by oil refining operations.

F. Chile

The legislation currently in force is of recent origin and general character (18). It states that gases, vapors, smoke, dust, emanations or contaminants of any kind in any factory or work place must be trapped or removed so as to prevent danger, damage or nuisance to a neighborhood. Combustion equipment which consumes solid or liquid fuel and is used for central heating or the production of hot water must be approved by the National Health Service, and municipal authorization of construction and modification of buildings must be subject to National Health Service air purity requirements. Persons engaged in the operation of combustion equipment must hold a certificate of competence from the National Health Service. The outdoor burning of leaves and rubbish in urban areas, and the operation of vehicles which emit visible smoke, are prohibited.

The National Health Service is responsible for the investigation of the nature and effect of contaminants; for laying down maximum permissible concentrators of contaminants; for establishing official methods of analysis; for specifying protective devices and measures; for approving the designs and plans for such devices; for setting time limits for plant modification; for approving plant designs and installations, for issuing certificates where required; for approving the operation of incinerators and in general ensuring that the legislation is observed.

G. Czechoslovakia

In 1966 an Order of the Minister of Health was made dealing with environmental sanitation and the establishment and protection of healthy living conditions (19). It is concerned with the prevention of contamination of the atmosphere, water, and the soil. Part 2, Chapter 1, Division 1 deals with air conditions and treats separately residential environments and areas for balneotherapeutic, climatotherapeutic, educational, and recreational purposes.

The former must be protected against contamination by dust, ashes, smoke, gases, vapors, odors and other substances dangerous to health. Fundamental hygiene requirements and maximum permissible concentration of pollutants are to be determined by the Chief Government Hygienist and made the subject of Ministerial directions. More detailed standards are to be determined for each locality (taking local conditions into account) by the regional and district hygienists. Plans and designs for the construction, modification or reconstruction of factories, installations and means of transport must provide for compliance with the standards required. Existing factories must be brought within these standards of operation. Section 3[3] provides widely that in all other activities every person concerned must take all necessary measures to ensure that the air is not affected in an objectionable manner.

Areas used for health or educational purposes are dealt with under Section 4 and are given special attention in addition to the measures prescribed in the preceding sections. New sources of pollution cannot be established and existing sources must be rapidly removed. Health protection zones may be established around factories and installations if the prescribed measures are insufficient to prevent all pollution. These zones are to be decided upon by the regional planning agencies and the duty of establishing them is upon the operator of the factory or installation.

In 1967 a law was passed concerned with measures to prevent air pollution (20). It deals mainly with fines in connection with offenses, and in an annex provides maximum permissible limits of discharges of flyash, sulphur dioxide and other specified harmful substances from any source.

H. Eire

Until 1962 the legislation in force, appropriate to the control of air pollution, dated from before the Constitution of the State in 1921. It was found to be inadequate to deal with the rapid postwar industrialization of Eire and in 1962 general enabling legislation was passed as part of the existing Local Government (Sanitary Services) Act (21). It is remarkable for its width and flexibility. The Minister for Local Government is given power to enact regulations controlling sources of atmospheric pollution; regulating the establishment and operation of trades, chemical and other works and processes which are potential sources of pollution; specifying maximum permissible concentrations of particular pollutants; regulating measurement and investigation of pollutants; regulating sources of radioactive pollution; specifying particular controls of air pollution for particular areas; licensing persons engaged in specific works or processes likely to cause pollution; licensing premises which discharge pollutants, prohibiting discharges from unlicensed premises; and cancelling or suspending such licences. Authorized officers are to administer these regulations and have power to enter premises for investigations.

I. France

Various decrees and prefectorial ordinances have been made for the control of air pollution since the early years of this century, although it was not until 1932 that a law dealing with air pollution generally and applying to the country as a whole was promulgated. In fact, industrial establishments are still classified according to a decree of 1917. This classification distinguishes between establishments which must be located well outside urban areas, those which may be authorized to operate within such areas, and those which do not require authorization for their site.

In 1961 a general enabling law was passed to cover the prevention of atmospheric pollution and odors (22). It provides that all buildings, establishments and vehicles shall be constructed and operated so as to prevent air pollution and odors which are a public nuisance and a danger to health. There is power under the Act to make provisions regarding conditions where the discharge of smoke, dust or poisonous gases or of corrosive, malodourous or radioactive substances may be regulated or prohibited. Certain establishments are classified and there is provision for establishments not listed to be dealt with by the Prefect, Mayor, and Department Board of Public Health jointly, if they present a serious danger or discomfort to the neighborhood. The Act was designed to be implemented by subsequent decrees, and this process has begun with a decree of September 1963 (23).

Part I of this decree deals with measures which are to apply to the country as a whole. The Ministers for Industry and Public Health and Population may lay down in orders specification for fuel-burning equipment which must be compiled with before its importation, manufacture or sale within France. It provides that appointed inspectors are to have access to such equipment for investigation. Where an installation is a source of pollution which threatens public health, the Prefect may order appropriate measures of abatement to be taken within a definite time. If these orders are not complied with summary procedure may be taken by the Prefect.

Part II deals with special protection areas. These areas are to be delineated by joint orders of the Ministers for Public Health and Population, the Interior, Industry, and Construction, made on the proposal of the Prefect after consultation with the Departmental Health Council. The extent of the area should depend on the size and composition of the population, local climate and types of pollution; and the extent of the areas may be varied from time to time. Within these areas, all establishments, including domestic and public, must comply with regulations for air purity made by the Ministers for Industry and Public Health and Population. Provision is made for existing establishments to be brought within the prescribed standards.

J. German Democratic Republic

In June 1968 the Minister for Health published an order dealing with the control of air pollution (24). It applies to industrial plant, residential premises and forms of transport which cause pollution. The term "polluting materials" is defined as being any solid, liquid or gaseous material which alters the natural composition of the air. The regulations must be taken into account by the relevant authorities when granting permits for the construction and operation of buildings and plants, and, where necessary, such buildings or plants must be equipped with devices to enable them to comply with the regulations. Maximum permissible standards of emission are prescribed and reference is made to the establishment of uniform standards of air quality measurement. Where the installation of air cleaning equipment is insufficient to prevent emissions exceeding the allowed limits then other steps may be ordered. These may involve prescription of chimney heights, regulation of fuels used or reduction of the size of the operation. Provision is made for exemptions or variations in particular cases.

K. Germany, Federal Republic Of,

Legislative action has been taken on both a federal and a state level. In 1959 the Federal Act to amend the Industrial Code and to complete the Civil

Code came into force. Permits must be obtained for the construction of plants likely to cause a nuisance or a threat to health, and a list of the plants affected is given in an Ordinance of 1960. The law of 1965 (25) concerning purity of the air directs the federal Minister for Health to undertake measurements of the nature and quantity of gases polluting the atmosphere. He is to assess the state and development of air pollution in the Republic and provide a basis for steps to be taken to prevent and abate such pollution. Recommendations may be made to the competent state authorities, and these authorities must report annually to the Bundestag and Bundesrat on the state and development of atmospheric pollution within their jurisdictions.

The State of North Rhine—Westphalia passed legislation concerning emissions of pollutants in 1962 (26). This State contains most of the heavily industrialized districts of the Republic and its legislative lead has been followed fairly closely by the States of Baden—Wurttemberg, Bavaria, Rhineland—Pfalz and Lower Saxony.

The 1962 law is nontechnical in character and covers establishments, whether or not industrial, which give rise to air pollution, noise or vibration. "Air pollution" is defined as "any modification of the natural composition of the air by the introduction of smoke, soot, dust, gas, vapors or odors." Any person operating such establishments is required to install, operate, and maintain the plant in such a fashion as to protect the neighborhood from sanitary nuisances. The State Government may by order implement the law by specifying *inter alia*, requirements for the construction and operation of establishments, the maximum permissible levels of pollution and regulating the use of fuels.

L. Honduras

Air purity legislation in Honduras is still of a rather general nature. Chapter III of the Sanitary Code deals with environmental hygiene (27). Under Section 77 the General Directorate of Public Health is required to take all necessary measures and issue appropriate orders to protect the public from nuisances or health risks arising from, *inter alia*, toxic gases, smoke, dust or noxious emanations of any type. The setting up of industrial establishments outside areas specified by the competent authorities, whereby their operation constitutes a source of "insalubrity or nuisance" to the public is prohibited by Section 78.

M. Italy

The general law of July 1966 (28) sets out to provide for the control of atmospheric pollution. Chapter I contains the general provisions and the law is stated to apply to the operation of solid-fuel or liquid-fuel fired thermal instalations, and the operation of industrial instalations and motor vehicles giving rise to the discharge of smoke, dust, gas, and odors of any kind causing a deterioration in the normal sanitary condition of the air and thus endangering public health or damaging property. The country is divided into two zones for the purpose of prevention of pollution; the compass of the zones being based on population criteria and severity of pollution levels. A Central Board for the Control of Air Pollution is established in the Ministry of Health with the functions of investigating air pollution problems, advising on all relevant questions submitted to it, and encouraging study and research. A Regional Committee for the Control of Air Pollution attached to the office of the provincial medical officer is to be established in the chief town of every region affected by the law. The regional Committees have functions similar to those of the Central Board.

Chapter II of the law sets out the requirements for the construction and operation of thermal installations located in either zone A or B, and Chapter III deals with the regulation of the use of fuels. Chapter IV provides for the operation and surveillance of installations affected by the law. Industrial establishments are covered by Chapter V. They must possess, in conformity with regulations made under the law, equipment capable of restricting as much as possible the emission of pollutants. The Regional Committees have the duty of inspecting and enforcing the provisions relevant to industrial plants. Under the provisions of Chapter VI motor vehicles are prohibited from producing fumes containing pollutants in amounts greater than those to be laid down by regulation. The Minister of Health, in agreement with the Ministers of the Interior, Transport and Civil Aviation, Industry and Commerce, and Labor and Social Insurance, may require the use of pollution control devices on all vehicles powered by internal combustion engines. The final provisions contained in Chapter VII require regulations for the implementation of this law to be promulgated by a Decree of the Head of State within 6 months of its coming into force. Regulations have appeared in various forms since the passing of the law. They cover *inter alia* fuel-burning installations which are not used in industrial production (29) and the delineation of the zones referred to in the law itself (30).

N. Jamaica

Although a relatively nonindustrialized country, Jamaica introduced legislative control of some sources of air pollution in 1961 (31). The Clean Air Law is administered by the Central Board of Health and implemented by officers appointed by the Minister. It provides that an owner of industrial works likely to discharge smoke, fumes, gases or dust must use the best practicable methods either to prevent the discharge or to render it harmless or inoffensive. The word "practicable" is defined to mean "reasonably practicable having regard *inter alia* to local conditions and circumstances, to the financial implications and to the current state of technical knowledge." The Central Board of Health may order further control measures if it is not satisfied that steps already adopted by the owner of any works are adequate, or if no such steps have been taken at all. Provision is also made for the owner of any premises to appeal to a judge in chambers against any order made in respect of his establishment.

O. Japan

In 1962 an Air Pollution Control Act was passed which is of a general enabling nature (32). It deals with emission of soot and smoke, and any other specially designated harmful materials, from industrial installations. These include sulphurous acid and sulphur dioxide and such other materials as may be listed in Cabinet orders. Any person intending to establish an installation which will emit soot or smoke must report to the Provincial Governor indicating the type of installation, its mode of function and proposed methods for pollution control. Existing installations must comply with this requirement within 30 days of the law coming into force, and the Provincial Governor may order any necessary change or modification in proposed pollution control methods. The Ministers for Welfare and Trade and Industry are to issue rules covering permissible discharge levels of smoke and soot and the measurement and record of pollution densities. Provision is made for emergency measures to be taken if it becomes necessary to avert an immediate threat to health.

P. Malta

In 1967 a Clean Air Act (33) was passed which provides for the establishment of a Clean Air Board with the duty to review the progress of air pollution control and abatement, to obtain advice concerning problems related to control and to advise the Minister on methods of air quality management. The Minister may make regulations concerning the taking of samples and measurements, the keeping of records, and measures for the control of any chimney smoke or other matter which may cause pollution or be prejudicial to health or constitute a nuisance. The Act specifically prohibits the emission of dark smoke from the chimney of any building and also applies to vessels while in Maltese waters. It requires the submission to the Board of plans for any chimney of a building which is used as a residence shop or office, and which is likely to cause pollution.

Q. Netherlands

In 1963 an Air Pollution Advisory Council (hereinafter referred to as the A.P.A.C.) was created with headquarters at the Hague (34). It is charged with the duty to examine problems related to the control of air pollution and to advise the Minister for Social Welfare and Public Health as to possible measures to be taken. However, under the Public Nuisance Law of 1952, a permit is required for establishments which may cause danger, damage or nuisance to their surroundings, and these are specified by regulation. Until more specific and concrete measures are taken by the A.P.A.C. the nuisance regulations form the basis of air pollution control legislation.

R. New Zealand

Air pollution control legislation in New Zealand is embodied in Part V of the Public Health Act 1956 (35). It deals mainly with chemical works and provides for their inspection in order to determine the amount of discharge of smoke, fumes, gases or dust. Tests, sampling, and analysis may be carried out to this end. The occupier of any chemical works must use the best practicable means to prevent the escape of noxious or offensive gases (as set out in Schedule 5 to the Act) or to render such discharges harmless. The "best practicable means" includes the installation, maintenance and operation of an air cleaning apparatus. The competent inspectors are to determine "best practicable means" and an aggrieved party may appeal in writing to the Minister who shall constitute a Board of Appeal.

Occupiers of fertilizer or sulphuric acid works are directed to do all things reasonably necessary to ensure that the acid gases of sulphur, or of sulphur and nitrogen are condensed as may be prescribed by regulation. Regulations governing emissions from these works were made in 1957. Smoke restriction regulations were passed in 1964. They do not apply to private dwellings. In August 1970 a Committee of the Board of Health published a report on air pollution problems in New Zealand. It made sweeping recommendations for a new legislative approach. If the recommendations are implemented they will establish a comprehensive scheme which would control through a system of licensing certain processes, and would also provide for the general regulation of all sources of air pollution both domestic and commercial. It would be administered and enforced through the co-operative efforts of a central agency and local municipal officers. See New Zealand Board Health Report Series., 1970, 15.

S. Philippines

An Act of June 1964 created the National Water and Air Pollution Control Commission (hereinafter referred to as the N.W.A.P.C.C.) with headquarters in Manila (36).   The Act defines pollution broadly but comprehensively, and specifically prohibits the direct or indirect discharge into the atmosphere of organic or inorganic matter as any substance in gaseous or liquid form liable to cause pollution.   A permit from the relevant city or district engineer is required for *inter alia* the construction, installation, modification or operation of any industrial or commercial establishment likely to cause an increase in the discharge of wastes into the water or atmosphere or to alter the physical, chemical or biological properties of the atmosphere in any manner unauthorized by law.   A permit must also be obtained for the construction or use of any new outlet for the discharge of gaseous or liquid waste into the water or atmosphere of the Philippines.

The N.W.A.P.C.C. is authorized to determine the presence of air and water pollution and to promulgate any necessary rules and regulations, to make orders requiring the discontinuance of activities causing pollution, and to issue, modify or revoke permits for the discharge of waste.   Some other responsibilities of the N.W.A.P.C.C. are to encourage voluntary cooperation by the public, municipalities and industry in proper conservation of air and water, to serve as an arbitrator for determining reparations for damage and losses caused by pollution, to prepare and develop a comprehensive plan for the abatement and future prevention of pollution, to issue standards, rules and regulations for sewage and industrial waste disposal systems, and to collect and disseminate information on air and water pollution, its abatement and control (37).

T. Poland

In 1966 a law concerning the protection of the atmosphere against pollution was passed (38).   It is of a general nature, covers all sources of pollution and delegates authority to control air pollution to several authorities.   Its stated aim is to protect against air pollution and to abate existing sources of pollution.   "Atmospheric pollution" is defined as being any solid, liquid or gaseous substance in an amount which exceeds the permitted maximum concentration.   The Council of Ministers has power to lay down such standards and to provide for exemptions from them where economically justified.

Every undertaking which causes or may cause pollution must construct or install and operate the equipment necessary to achieve the permissible standards of emission.   This requirement applies to state or other socialized establishments, or a corporate body which carries out the industrial activities of a natural person.   Designs and plans for the construction of these undertakings must be approved by a competent agency.   The Chairman of the Central Bureau for Water Resources Management designates these agencies. Provision is made for the establishment of "protective zones" around undertakings liable to cause pollution as determined by the Council of Ministers by

Order. Officers of the competent agencies are able to enter premises for the purposes of testing, measuring and sampling and to check the operation and use of equipment. Where a discharge is found to be in excess of the maximum permitted, the agency responsible may call on the undertaking to remedy the situation within a specified time, and on expiry of this time without effective action being taken, may order the cessation of the undertaking. The Council of Ministers may prohibit the use in a given area of certain fuels, processes or raw materials or the use of certain fuels by motor vehicles if they are presenting a threat to health, or in circumstances where unfavorable weather conditions pose a threat to health. There is also provision for bringing existing undertakings within the air pollution standards prescribed.

Several authorities have issued decrees implementing this basic law. Illustrative of the procedure is the decree of the Council of Ministers of March 23, 1967, authorizing the Chairman of the Central Water Utilization Bureau to establish the width of protective zones around sources of pollution (39). The Chairman of the Bureau himself issued a decree prescribing the widths of these protective zones on May 30, 1967 (40).

U. South Africa

The Atmospheric Pollution Prevention Act of 1965 established the National Air Pollution Advisory Committee and Appeal Board as appointed by the Minister of Health (41). Its functions are to advise the Minister on control, abatement and regulation of pollution, to stimulate public interest and to consider and advise on any relevant matter.

The Minister may declare controlled areas where a certificate of authorization is required to erect, alter or carry on a scheduled process. Smoke pollution of a Ringleman color darker than prescribed by regulation is prohibited; although these provisions apply only to specified areas and do not cover private dwellings. Local authorities have wide regulatory powers regarding smoke emission, Part V of the Act deals with regulations which may be made prohibiting vehicles emitting fumes of a certain darkness, and local authorities may authorize officers to implement them. Dust pollution is covered by Part IV, which provides for the prohibition in certain areas of a process which causes a dust nuisance.

V. Switzerland

In 1960 a Federal Commission for Atmospheric Hygiene was established as an advisory body of the Department of the Interior (42). It is charged with collecting and utilizing observations and measurements of types of air pollutants, their effects and possible protective measures, and suggesting legislative steps to be taken.

W. United Kingdom

Atmospheric pollution is covered by two complementary acts. Pollution arising from chemical processes is dealt with under the Alkali etc. Works Regulation Act of 1906, and the emission of smoke, dust and grit from any building, chimney or plant is controlled by the Clean Air Act of 1956.

The Alkali Act is enforced by specially appointed inspectors responsible to the Minister of Housing and Local Government and the Secretary of State for Scotland. These inspectors have the task, in particular, of deciding what constitutes "best practicable means" for prevention of gaseous discharges with respect to any given process. The Act basically requires that certain processes listed in a schedule to the Act may operate only with a certificate of registration, renewable annually. The certificates are granted subject to the process being equipped to operate in conformity with the Act. The owner of any process must use the "best practicable means" for preventing the discharge of noxious or offensive gases and for rendering such gases, where discharged, harmless and inoffensive.

The Clean Air Act provides the basis for the making of regulations by the Minister of Housing and Local Government in respect of England and Wales with regard to the technical details of control of emissions of grit, dust and smoke. It provides for the approval by the Local Authority of plans for new plants, and gives these Authorities power to set up smoke control areas where any emission of smoke is an offense, whether or not it is discharged from a domestic source. However, certain buildings may be exempted and the Minister may suspend or relax the operations of these provisions. Burdens to householders involved in compliance with the provisions of the smokeless zones are reduced by a grant of 70 % of the cost of conversion from solid-fuel burners. Provision is made for the regulation of chimney heights and mine refuse, and the smoke control provisions of the Act are applicable to railway engines and vessels within the territorial waters of the United Kingdom. The Act does not apply to motor vehicles which are dealt with under the Road Traffic Act of 1930–1947. One Clean Air Council with the duty to conduct research into problems of pollution and give advice regarding measures for control is established for England and Wales and one for Scotland.

Since the passage of the Clean Air Act, certain regulations have been made for its enforcement. Of these, the most important are the Dark Smoke (Permitted Periods) Regulations of 1958. They define "black smoke" as being darker than Ringleman Chart Shade No. 4, and also give the permitted periods for the emission of dark smoke from a chimney (43).

X. United States of America

In the United States specifically anti-air pollution legislation has been left largely in the hands of the state and municipal authorities. Because of the multiplicity of these authorities and the laws and regulations enacted, it is impossible to consider them in detail in this survey. Therefore, the reader should refer to the Digests of State Air Pollution Laws issued by the Public Health Service of the U.S. Department of Health Education and Welfare and Vol. 14, International Digest of Health Legislation page 187, at page 122.

The local government bodies in the U.S.A. counties which are most affected by air pollution have taken the lead in control programs. In 1963 the average *per capita* expenditure by a local agency was 5.8 times the average state expenditure. All the large effective local or regional control agencies, of which Los Angeles, Chicago and the San Francisco Bay Area are the best known, have administrative, engineering, enforcement, and technical services and public information centers staffed with professionals in each field. Some of the most spectacular results in abatement have taken place in areas controlled by local agencies.

Recently, however, states have become more involved in the legislative program of air pollution control. In 1963 only 17 states had a realistic and active program of control. In 1967 this number had increased to 33. However, only 8 of these states have operating budgets larger than the largest local agency within the state. The total of local expenditures in California in 1967 was more than twice the state budget. In the United States the states are relative beginners in the field of air pollution control, although they are strongly supported by the federal legislation and their involvement is rapidly increasing.

The federal program itself dates from 1955 when legislation authorizing research and technical assistance to state and local agencies came into force. Congress then established the policy that state and local governments have a fundamental responsibility for dealing with community air problems. The role of the federal government was seen as one which provided leadership and support. This policy has not changed radically since 1955 although the ways in which the federal government makes its aid available have developed considerably. In 1960 it entered the field of automobile emissions by instituting a special study of the problems involved to pave the way for national control action and in 1965 the Clean Air Act was amended to allow for the federal establishment of standards of motor vehicle pollution. The Clean Air Act of 1963 was itself a major step in federal control action. Since 1963, amendments to the Clean Air Act have broadened federal activity in research and increased the purposes for which grants can be made directly to the state or local agencies. The latest phase of federal involvement is embodied in the 1967 Air Quality Act, which amended the Clean Air Act in a number of significant respects. Eight atmospheric areas covering 48 states were defined as a preliminary step towards the designation of air quality regions.   These regions are defined independently of state and county borders and are based on atmospheric, industrial, topographical, and other pertinent conditions which suggest that a group of communities should be treated as a unit for the purpose of setting up and implementing air quality standards.   Another important amendment authorizes the Department of Health, Education, and Welfare* (H.E.W.) to establish standards if a state fails to act or if a state action is considered inadequate, and to enforce such standards by court action where the pollution can be shown to affect another state.   Machinery is also set up to allow immediate court action by H.E.W. to stop pollution in emergency situations endangering health, regardless of the economic and technological feasibility of pollution control.

The Air Quality Act has been criticized as not being "the strong decisive instrument needed to bring about an immediate significant reduction in air pollution levels" (44).   Furthermore, despite local government's record of achievement in air pollution control, the Air Quality Act is state-oriented and represents a departure from the past policy of the federal government in this respect.   Although some states are actively involved in control programs, the most efficient, well-organized and developed schemes are run by local agencies.   Thus, the problem raised by the Act is how to completely reorientate the structure of legislative control and enforcement without losing the race against pollution.

Y. United Soviet Socialist Republic

Although the U.S.S.R. was one of the first countries to introduce specifically anti-air pollution legislation, materials available do not reveal a great deal of development at a legislative level.   Control by departmental directions may account for this scarcity of information.

Under an order of June 1949 (45), electric power stations or houses may not operate without equipment for the absorption of dust and ashes, and authority is required for the establishment of likely sources of pollution. Metal works for nonferrous metals may not operate without apparatus to absorb dust and matter containing arsenic or fluorine.   Similar conditions are applicable to factories for the distillation of coal regarding sulphuretted hydrogen and sulphurous gases, factories for the treatment of iron regarding blast furnace gases and factories using dissolvents without installing a method for waste recovery.

A list of maximum permissible concentrations of harmful substances in the atmosphere of inhabited localities was approved by the Chief State Inspector of the U.S.S.R. in 1963.   Ambient air quality is protected by a

---

* Since December 1970 this authority has been vested in the Environmental Protection Agency (E.P.A.) which includes the Air Pollution Control Office (A.P.C.O.).

series of compulsory measures generally administered by the sanitary-epidemiological services of the U.S.S.R. and the constituent republics.   No power station or industrial establishment may be planned or operated unless measures have been taken to control the emission of pollutants so as to comply with the relevant air quality standards.   The sanitary-epidemiological services must authorize the operation of any plant or installation which is a potential source of pollution, and plans for the construction or expansion of industrial establishments cannot be approved unless provision has been made for the installation of air cleaning equipment.   New establishments which are technically incapable of complying with these requirements can be located only at regulated distances from inhabited areas, parks, and forest belts as prescribed by the sanitary-epidemiological authorities.   All other industrial establishments must be separated from residential areas by a protective zone as prescribed by the State Building Construction Committee.

Yearly national economy plans make special provision for the establishment of effluent control installations in industrial establishments already in operation.   Air cleaning installations are subject to surveillance by an interministerial organization, and there is a continuous surveillance of the ambient air quality by the sanitary-epidemiological services under the regulations of the Council of Ministers of 1963.   Registration with the relevant sanitary-epidemiological station is required of all industrial establishments which involve the pollution of the air of inhabited localities.   These services also have general duties of research into and, planning and administration of air sanitation programs.   Provision is also made for emergency measures when a source of pollution is uncontrollable and dangerous to the locality.

Z. Yugoslavia

The basic law concerning the protection of the air against pollution was promulgated in June 1965 (46).   The general provisions state that the law deals with the surveillance of establishments giving rise to pollution, and lays down measures for the abatement of such pollution.   "Air pollution" is defined as being the discharge of gases, vapors, smoke, dust, and radioactive materials and other substances in quantities liable to have a harmful effect on population or property.   The sources of air pollution covered by the law are described and provision is made for future application to other sources such as private dwellings.

Measures to be taken under the law include the approval of plans and construction of plants and the creation of health protection zones to protect residential areas.   The specifications for these zones and the classification of establishments likely to cause air pollution are laid down by certain federal authorities.   Town plans must comply with the requirements of this law under the supervision of the local agency responsible for sanitary inspection, and every commune must arrange for the testing of air quality within its own area, although several communes may arrange to do so jointly. Each commune, through its competent agencies, is made responsible for the organization and implementation of such measures as may be prescribed, as well as for the approval of institutions responsible for the surveillance of plants, organization of training courses for the staff of the relevant agencies, and assisting in the design of control equipment. The Federation, through its relevant agencies and the Federal Institute for Health Protection, is responsible for coordinating research into air pollution problems and the setting of air quality standards.

The communal agencies responsible for surveillance may prohibit the construction or use of establishments, installations, vehicles, and equipment where protection from pollution cannot be guaranteed. They may also regulate trade in control devices and their installation and maintenance.

The Federal Secretary for Health and Social Affairs on the proposal of the Federal Institute for Health Protection may make regulations covering the location and construction of establishments and installations, the use of equipment, maximum permissible emissions, conditions of surveillance, the keeping of records and methods for determining air purity. In March 1966 an Order of the Federal Secretary for Health and Social Affairs was made covering the notification of establishments likely to cause atmospheric pollution (47).

V. GUIDELINES FOR LEGISLATIVE PROGRAMS

The Expert Committee on Environmental Sanitation, which studied the question of air pollution in 1958, suggested that legislation should be directed towards:

1. the control of sources of pollution by specifying the types of industrial and other processes which should operate under supervision by control authorities, and the types of emissions which should be kept to a minimum value;

2. the institution of town planning practices in which due attention is given to the planning and zoning of industrial sites for the purpose of reducing air pollution, providing always that such action does not make the conduct of the industry prohibitively costly or even impossible;

3. the provision of regulations governing the types of fuel to be burned in installations where combustion emissions are not otherwise controllable (48).

Clearly the precise method of control and abatement chosen by any state or country will depend upon its political and administrative structure, its economic progress and degree of industrialization. The following points are aimed at identifying in general terms a number of features which are desirable to include in any legislative scheme aimed at air quality management, regardless of the actual mode of implementation which is adopted.

A. General Sections

A legislative program to control and abate air pollution should be flexible and yet sufficiently certain in its terms to allow easy application and enforcement. A clear statement of the policy and aims of the legislation facilitates its interpretation by the officers charged with its administration and the courts which deal with its enforcement. A definition section is essential in any legislation which contains terms which are to bear a special meaning. However, exhaustive definitions are generally unnecessary except with reference to these special terms. An excessively broad and therefore vague definition of any term should be avoided.

A good example of a definition of "pollution" which is both comprehensive and yet specific enough to avoid difficulties of intepretation may be found in the Philippine legislation of 1964, referred to earlier. In that Act, which deals with conservation of both air and water quality, "pollution" is defined as being such alteration of the physical, chemical or biological properties of water and or the atmospheric air, or any such discharge of any liquid, solid or gaseous substance into the water or air as will or is likely to create or render such water or air harmful or detrimental or injurious to public health, safety or welfare or to domestic, commercial, agricultural, industrial, recreational or other legitimate uses or to livestock, wild animals, birds, fish or other aquatic life.

B. Administration and Enforcement

The most desirable approach for any legislation of this nature is to regulate through the lowest level of government which is consistent with the most practically efficient control of pollution. The most effective control might be achieved on the basis of air basins rather than political subdivisions. An air basin is an air mass which conforms to the same air flow pattern and is subject to essentially the same pollution. Thus, local governmental authorities such as municipal or city councils may not be the most practical administrative unit through which to set about control and abatement of air pollution.

The creation of a separate and autonomous administrative agency is an essential feature of a successful scheme. Such an agency should be constituted so as to take into account the variety of interests which may be affected by air quality control legislation, and it seems advisable to provide for representation of these interests through an advisory council to the agency. The agency should be empowered to make regulations for the implementation of the basic legislative policy and to administer and enforce such regulations. It should be able to conduct, coordinate, and assimilate research, to provide for the training and appointment of properly qualified staff, to conduct programs for the education of the public and to enter into abatement schemes jointly with other bodies should the need arise.

Procedure for appeals from the decisions of the agency should be incorporated in any legislation. The necessity for variations from regulations stems from the fact that installations cannot always both comply with the stated requirements and continue economic operation. However, variations should not be made where the hardships occasioned to the public are greater than those suffered by the individual applicant. If possible, procedure for appeal should approximate to that of normal court practice and provision should be made for further appeal to a court of competent jurisdiction.

Strict and proper enforcement of the scheme is essential to the ultimate efficacy of the legislation. The penalties imposed for violations of the statutory requirements or regulations should be realistically calculated, and the impartiality of the enforcement officers should be maintained as an essential requirement for their appointment.

C. Regulations

Regulations should be made as clear as possible so as to encourage obedience and facilitate enforcement. The scheme should have as wide as possible an application. It should not be limited to apply only to industrial or commercial plants, but should be expressed to cover all public and private activities and instrumentalities which are potential sources of pollution. The agency should have power to widen the application of its control to sources of pollution which may develop in the future, and to remove from that control apparent sources of pollution which are proved to be harmless in fact. Automotive pollution should be the subject of special treatment under the legislation, as it appears to be one of the largest single sources of air pollution in cities but is not so easily policed as stationary sources.

The permit-oriented approach to regulation of air pollution at its source is perhaps one of the best documented methods of successful control and abatement. This technique of control requires the submission of plans and designs to the relevant agency for approval before a permit to construct or operate a certain plant is obtained. This method requires a high degree of administrative responsibility but has the advantages of supplying a source of revenue to the agency through permit fees and making abatement a simple matter of withdrawing or qualifying the permit. Regular inspection of premises to ensure continued compliance with the conditions of the permit is also necessary.

Some schemes have been based on a system of registration of certain plants or installations. This method allows the relevant agency to set about abatement with reference to the location and nature of the potential sources of pollution. A combination of the permit and registration schemes would seem to be an efficient method of control. Conditions which might be attached to the granting of a permit are requirements for the installation, use and proper maintenance of air cleaning equipment where necessary to reduce emission levels in order to comply with prescribed standards. It may also be necessary to regulate the use of types of fuel in combustion processes, and to make provision for the establishment of pollution-free zones along lines similar to the "smokeless zones" of the United Kingdom Clean Air Act of 1956. The creation of buffer zones of park land around residential and recreational areas would not substantially reduce pollution but would probably reduce its depressing effect.

It is not essential, although it may be a good thing, to make provision for payment of compensation to victims of damage caused through pollution which has been the subject of an action by the enforcement agency. Although this would overlap to some extent with the role of the common law actions discussed earlier, it would provide an added deterrent to pollution, and avoid some of the procedural drawbacks of forcing victims to bring separate common law actions themselves.

The value of encouraging cooperation from the public, industry, and commerce should not be underestimated. Although no real evidence is available which supports any particular method of encouragement above others, the possibility of tax concessions or deductions for those who must spend money in order to comply with requirements should be investigated. The research facilities of the enforcement agency and its advice and equipment ought to be available for use by interested parties.

References

1. (1954) 27 *Southern California Law Review* 347, 351.
2. Reingold, P. D., (1966–1967) 33 *Brooklyn Law Review* 17, 28.
3. For a more detailed discussion of the problems posed by private actions for compensation for damage caused by air pollution see Wolf, (1968) 5 *Trial Lawyer's Quarterly* 22. Liability for negligence under the Continental Codes is treated at some depth in Ryan, *An Introduction to the Civil Law*, Law Book Company (Melb.), 1962, and under the common law in Prosser, "*Handbook of the Law of Torts.*"
4. See for example the Code of Alabama, Recompiled 1958, Title 22, ss. 75 and 76.
5. Pannam, C. L., "Recovery of Unconstitutional Taxes" (1964) 42 *Texas Law Review* 777.
6. Public Health Service Publication No. 1549 issued by the United States Department of Health, Education and Welfare.

6a. The law is stated as at 1 December, 1969.

6b. Parliament of the Commonwealth of Australia, 1969. Parliamentary Paper No. 91.

7. Act No. 6125 of 1957 (Acts of Parliament, 1957 Part II, pp. 1008–1015) see 10 *I.D.H.L.* 429.

8. Act No. 69 of 1961 (Statutes of New South Wales 1961, pp. 475–500) see 15 *I.D.H.L.* 7.

9. Moniteur Belge 14 January 1965, No. 9, pp. 345–347, see 16 *I.D.H.L.* 677.

10. Decree-Law No. 303 of 20 February 1967 (Diario Official, Section 1, Part 1, 28 February 1967, No. 40, pp. 2480–2481) see 19 *I.D.H.L.* 723.

11. Decree No. 728 of 24 October 1963 (D'rzhaven Vestnik, 29 October 1963, No. 84, pp. 1–2) see 16 *I.D.H.L.* 33.

12. Decree No. 45 of 24 September 1964 (D'rzhaven Vestnik, 9 October 1964, No. 80, pp. 1–3) see 16 *I.D.H.L.* 480.

13. Statutes of Ontario, 1958, Chapter 2, 8 pp.; See 10 *I.D.H.L.* 489.

14. Statutes of Nova Scotia 1960, Chapter 51, p. 216, see 12 *I.D.H.L.* 287.

15. The Manitoba Gazette 30 April 1960, pp. 667–668, see 12 *I.D.H.L.* 718.

16. Alberta Regulation No. 262 of 30 August 1961, see 13 *I.D.H.L.* 469.

17. Statutes of Saskatchewan 1965, Chapter 65, pp. 225–233, see 17 *I.D.H.L.* 310.

18. Decree No. 144 of 2 May 1961 (Diario Oficial de la Republica de Chile, 10 May 1961, No. 24947, pp. 978–979) see 14 *I.D.H.L.* 250.

19. Order No. 35 of 13 June 1966 (Sbirka Zakonu Ceskoslovensk Socialisticke Republicky, 29 June 1966, No. 17, pp. 189–198) see 18 *I.D.H.L.* 328.

20. Law No. 35 of 7 April 1967 (Sbirka Zakonu Ceskoslovensk Socialisticke Republicky, 18 April 1967, No. 13, pp. 118–124) see 19 *I.D.H.L.* 323.

21. The Local Government (Sanitary Services) Act 1962. (The Acts of the Oireachtas 1962, pp. 759–791) see 17 *I.D.H.L.* 904.

22. Law No. 61-842 of 2 August 1961 (Journal Officiel de la Republique Francaise, 3 August 1961, No. 181, pp. 7195–7197) see 13 *I.D.H.L.* 531.

23. Decree No. 63-963 of 17 September 1963 (Journal Officiel de la Republique Francaise, 21 September 1963, No. 222, pp. 8539–8540) see 15 *I.D.H.L.* 526.

24. Regulation of 28 June 1968 (Gesetzblatt der Deutschen Demokratischen Republik, Part II, 25 July 1968, No. 80, pp. 640–643).

25. Law of 17 May 1965 (Bundesgesetzblatt Part I, 22 May 1965, No. 21, pp. 413–415) see 17 *I.D.H.L.* 107.

26. Law of 30 April 1962 (Gesetz und Verordnungsblatt fur das Land Nordrhein-Westfalen, 15 May 1962, Part A, No. 31, pp. 225–227) see 13 *I.D.H.L.* 679.

27. Decree No. 75 of 14 November 1966 (La Gaceta, 5 January 1967, No. 19037, pp. 1–5, and 6 January, No. 19038, pp. 1–4) see 18 *I.D.H.L.* 728.

28. Law No. 615 of 13 July 1966 (Gazzetta Ufficiale della Repubblica Italiana, Part 1, 13 August 1966, No. 201, pp. 4091–4096) see 18 *I.D.H.L.* 667.

29. Decree of 24 October 1967 (Gazzetta Ufficiale della Repubblica Italiana, 9 January 1968, No. 6) see 20 *I.D.H.L.* 233.

30. Decree of 23 November 1967, (Gazzetta Ufficiale della Repubblica Italiana, 13 December 1967, No. 310).

31. Law No. 32 of 19 December 1961 (The Laws of Jamaica 1961, 1962) see 14 *I.D.H.L* 655.

32. No. 146 of 2 June 1962, see 16 *I.D.H.L.* 699.

33. Act No. XVIII of 1967. (Laws made by the Legislature during the year 1967 published by the Government of Malta and its Dependencies, 1968, Vol. 50, Part I, pp. A115–A122) see 20 *I.D.H.L.* 264.

34. Law No. 319 of 13 June 1963 (Staatsblad van het Koninkrijk der Nederlanden 1963, No. 319, pp. 1037–1038) see 17 *I.D.H.L.* 937.
35. Public Health Act 1956 (The Statutes of New Zealand 1956, 1957, Vol. II, pp. 959–1048) see 8 *I.D.H.L.* 665.
36. Republic Act No. 3931 of 18 June 1964 (Official Gazette 9 November 1964, Vol. 60, No. 45, pp. 7345–7352) see 18 *I.D.H.L.* 780.
37. Regulations setting maximum permissible standards of emissions have been passed. (Official Gazette 10 July 1967, Vol. 63, No. 28, pp. 5999–6018) see 20 *I.D.H.L.* 285.
38. Law of 21 April 1966 (Dziennik Ustaw Polskieg Rzeczypospoliteiz Ludowej, 29 April 1966, No. 14, Serial No. 87, pp. 113–115) see 18 *I.D.H.L.* 792.
39. (Dziennik Ustaw Polskieg Rzeczypospoliteiz, No. 15, Serial No. 66, pp. 95–96) see 19 *I.D.H.L.* 833.
40. Monitor Polski 15 June 1967, No. 32, Serial No. 152, pp. 341–345.
41. Government Gazette Extraordinary, 21 April 1965, No. 1089, pp. 24–72, see 17 *I.D.H.L.* 947.
42. Regulation of 19 January 1962 (Bulletin du Service federal de l'Hygiene publique, Supplement A, No. 2, 7 April 1962, pp. 25–26) see 14 *I.D.H.L.* 502.
43. A detailed discussion of the "Alkali" Act, the Clean Air Act and the "Dark Smoke" Regulations may be found in 14 *I.D.H.L.* at p. 212.
44. (1968) 33 *Law and Contemporary Problems* 292.
45. Order No. 431 of 14 June 1949 (summarized in Gigiena i Sanitariya, August 1949, No. 8, p. 62) see 2 *I.D.H.L.* 454.
46. Decree of 25 June 1965 (Sluzbeni list Socijalisticke Federativne Republicke Jugoslavijc, 7 July 1965, No. 30, Serial No. 525, pp. 1181–1183) see 17 *I.D.H.L.* 182.
47. Order No. 52-218/1 of 18 March 1966 (Sluzbeni list Socijalisticke Federativne Republicke Jugoslavijc, 6 April 1966, No. 14, Serial No. 183, pp. 287–288) see 19 *I.D.H.L.* 893.
48. World Health Organization, Tech. Rept. Series 157 (1958) at p. 18.

Control of Airborne Radioactive Effluents from Nuclear Reactors

G. W. Keilholtz
G. C. Battle

Oak Ridge National Laboratory
Oak Ridge, Tennessee

I. INTRODUCTION

One of the prerequisites to large-scale power production by the operation of nuclear reactors is a high degree of assurance that the containment systems and associated safety features of these reactors can preclude the uncontrolled release of fission products to the environment. In order to minimize the hazard to the public and to reactor personnel, this assurance must hold not only for normal operation and minor accidents but also for severe postulated accident conditions. The remarkable safety record (1) of nuclear reactors has been achieved through a very conservative approach to reactor design, licensing, and regulation. This approach has been implemented by exhaustive, continuing research and development work and great care in the design, testing, and operation of reactors and their components. Furthermore, full-scale commercial reactors of any type have always been preceded by very small experimental reactors and low-power prototypes. Such a safety record can be maintained during the coming decades only if the conservative philosophy and practice is continued and kept in pace with the exponential growth of electrical power production by nuclear reactors (2).

Approval by government regulatory agencies of the design, construction, location, and initial and continued operation of each nuclear reactor is governed by codes and regulations (3) that specify very low maximum permissible exposures of the public to radioactive isotopes under normal operating conditions and exposures not exceeding safe limits after improbable but conceivable accidents. These limits are set by the International Commission on Radiological Protection (4) and by similar committees and agencies in each country in which reactors are used. They are implemented by government regulations and regulatory agencies, with the aid and advice of specialists.

In the United States, for example, values of maximum permissible body burdens and maximum permissible concentrations of radionuclides in air and in water (or food) are recommended by the National Committee on Radiation Protection and Measurements (5) and the Federal Radiation Council. Values for some of the radionuclides most important in reactor operations are listed in Table I. The maximum permissible doses, both continuous and occupational, are established by "Atomic Energy Rules and Regulations," Title 10 of the *Code of Federal Regulations* (3). Under Title 10, Part 20 covers doses from normal reactor operations; Part 100 covers doses from postulated reactor accidents and all other criteria for reactor siting. The Division of Reactor Licensing, under the USAEC's Director of Regulations, is responsible for calculations and evaluations that are required to establish, for each reactor as a separate case, the maximum emissions of radioactive materials that can be permitted and the safety measures that must be taken in the design, construction, and operation of that reactor to ensure that these maximum levels are not exceeded. In this and other regulatory matters the USAEC is assisted by the Advisory Committee on Reactor Safeguards, a group of 15 independent advisors from industry, government, and universities who are responsible only to the United States Congress.

During the normal operation of power reactor plants, some of the gaseous and liquid radioactive fission products that accumulate in the fuel elements are released into the primary coolant system (described later) from leaking fuel elements. In addition, radioactive isotopes are produced by neutron activation in the coolant. The fission products and activation products that are most important in reactor operations are covered briefly in Section III. In some cases, solid fission products and fuel materials are leached from the fuel and released into the primary coolant. The release of these radioactive materials into the outside atmosphere is kept within allowable limits by coolant purification systems and high-efficiency air cleaning systems. These air cleaning systems are shown schematically in Section II and described in Section IV. Treatment of reactor off-gases during normal operations is discussed in Section V. Figure 1 gives an idea of the size of a reactor plant air cleaning system. Each housing contains, in banks 8 high by 4 wide, 32 moisture separators followed by 32 high-efficiency particulate air (HEPA) filters and 32 charcoal beds. This USAEC production reactor has 4 such housings on line plus one on standby. Air cleaning systems for power reactors are usually somewhat smaller (Sections V and VI).

TABLE I

Maximum Permissible Body Burdens and Maximum Permissible Concentrations in Air of Some of the Radionuclides Most Important in Reactor Operations[a]

Isotope	Form[b]	Critical organ of reference	Maximum permissible body burden (microcuries)	Maximum permissible concentration in air[c] (microcuries per cubic centimeter)	
				For 40-hr week (Occupational exposure)	For 168-hr week (Continuous exposure)
Tritium (Hydrogen-3)	HTO or T$_2$O (soluble)	Body tissue	10^3	5×10^{-6}	2×10^{-6}
	T$_2$ (immersion)	Total body	2×10^3	8×10^{-6}	3×10^{-6}
		Skin		2×10^{-3}	4×10^{-4}
Argon-41	^{41}A (immersion)	Total body		2×10^{-6}	4×10^{-7}
Krypton-85	(immersion)	Total body		1×10^{-5}	3×10^{-6}
Iodine-131	^{131}I (soluble)		0.7	9×10^{-9}	3×10^{-9}
	(insoluble)	GI(LLI)[d]		3×10^{-7}	1×10^{-7}
		Lung		3×10^{-7}	1×10^{-7}
Xenon-133	^{133}Xe (immersion)	Total body		1×10^{-5}	3×10^{-6}

[a] Based on Table 1 in ICRP Publication 2 (4) and Table 1 in NBS Handbook 69 (5). Those two tables, which are nearly identical, give values for about 240 radionuclides.

[b] Calculations for immersion are based on the dose a person would receive if he were surrounded by a hemispherical infinite cloud of radioactive gas. The term "immersion" is used in NBS-69; the term "submersion" is used in ICRP-2.

[c] Values for soluble tritium were taken from ICRP-2 only; the values in NBS-69 were corrected to the ICRP-2 values in Addendum 1 to NBS-69.

[d] Gastrointestinal tract (lower large intestine).

Fig. 1. Two filter housings installed on the roof of a production reactor building at the Savannah River Plant. These housings are mounted on railroad trucks for remote replacement (Courtesy of the Savannah River Plant).

Two types of power reactors, both moderated to convert the fast neutrons in the uranium-235 neutron emission spectrum to thermal (slower) neutrons for more efficient bombardment of uranium-235, are in wide and rapidly increasing use for power production. These are water-cooled and moderated, and gas-cooled (moderated with graphite). A third type, in the prototype stage of development, is the liquid-metal-cooled fast breeder reactor, fueled with plutonium-239 which indirectly produces more plutonium-239 by bombarding a surrounding "blanket" of uranium-238 with neutrons. Two other types that show considerable promise for the future are high-temperature helium-cooled reactors and molten-salt reactors. Prototypes of both of these have been operated successfully.   Also, both share with the liquid-cooled reactor the potential for use as breeder reactors.   The helium-cooled reactor may be operated with an unmoderated neutron spectrum as a fast breeder.   The molten-salt reactor, which is fueled with a circulating mixture of the molten fluoride salts of uranium and other metals, may be operated as a thermal breeder.

The core of a nuclear power reactor is a matrix of long, thin fuel elements, with the necessary control rods, instrumentation, and other components. The heat produced by nuclear reactions is used to heat a fluid, which is circulated between and around the fuel elements and from there to heat exchangers, steam generators, or turbines.   The fluid may be a liquid, such as water or molten sodium, or a gas, such as carbon dioxide or helium. This heat transfer fluid is conventionally called the primary coolant; "primary" because in most systems it serves to heat a secondary coolant and "coolant" because it cools the reactor core.   In the most direct type of system, the boiling-water reactor, the primary coolant water is allowed to boil in the core at high pressure, producing high-pressure steam which drives turbines directly, after which it is condensed and recirculated through the core.   In other types the primary coolant is passed through a steam generator or an intermediate heat exchanger.   In any case, high-temperature steam is generated and is used to power turbines that drive electrical generators.

The reactor components through which the primary coolant is circulated are known collectively as the reactor primary system, or primary coolant system.   This system is shown schematically in Figures 2 through 5 for 4 major types of power reactors.   (Section II contains schematic drawings showing power reactor containment systems and air cleaning systems.)   In all 4 types the primary system includes the reactor core (not shown), the reactor pressure vessel surrounding the core, the piping through which the coolant flows, and the necessary pumps (blowers in the case of the gas-cooled reactor).   The other components of the primary system vary with reactor type: turbines and turbine condensers in boiling-water reactors (Figure 2); steam generators and coolant purification systems in pressurized-water reactors (Figure 3); steam generators and filters in gas-cooled reactors (Figure 4); and intermediate heat exchangers and sodium purification system in liquid-metal-cooled reactors (Figure 5).

Regardless of the power of a reactor, licensing regulations require that its design includes provisions to ensure that the release of fission products and other radioactive materials to the environment will not exceed permissible limits.   This applies not only during normal operations but to radioactive materials that might be released as a result of any accident to that reactor, including the most severe possible accident, however improbable, that must be considered as a basis for the design of the reactor.

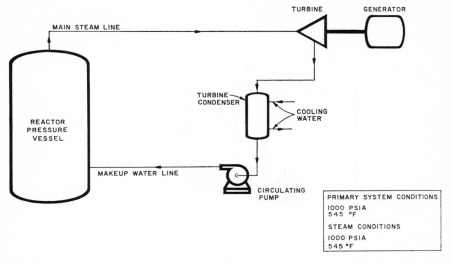

Fig. 2. Schematic of coolant system of boiling-water reactor. Steam produced in the core drives the turbine(s) directly (7).

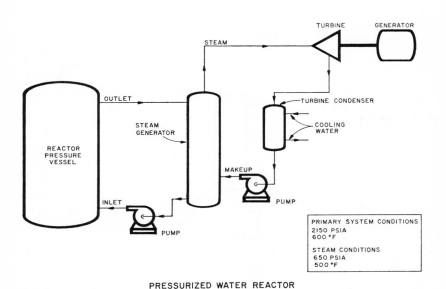

PRESSURIZED WATER REACTOR

Fig. 3. Schematic of coolant system of pressurized-water reactor. Water in the primary coolant system, which is under enough pressure to prevent boiling, generates steam in a secondary coolant system (7).

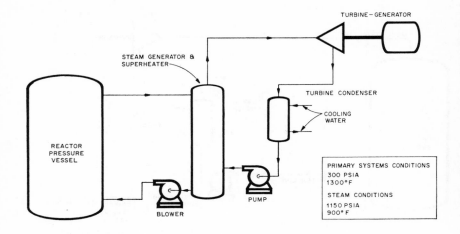

GAS COOLED REACTOR

Fig. 4. Schematic of coolant system of gas-cooled reactor. The gas in the primary coolant circuit generates steam in a secondary coolant system (7).

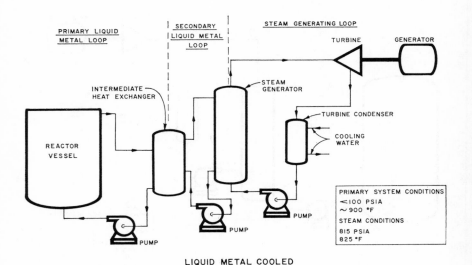

LIQUID METAL COOLED

Fig. 5. Schematic of coolant system of liquid-metal-cooled reactor. Radioactive sodium-24 is produced by exposure of the primary sodium coolant to neutrons. Therefore the primary coolant system is separated from the steam generator by a secondary sodium coolant system (7).

In the design and licensing of power reactor systems it is considered possible that under postualted accident conditions, gross amounts of radioactive materials might escape the confines of the reactor primary system. Therefore, the primary system is enclosed in one or more large structures specifically designed to contain airborne radioactive materials at pressures as high as about 60 psig in some designs. These structures and the equipment provided to enable them to withstand accident-produced pressures and temperatures are usually referred to as containment systems, or simply containment (Section II).

Most postulated design-basis accidents include rupture of the primary coolant system, gross cladding failure in many of the fuel elements, and partial or gross meltdown and vaporization of the fuel. Such an accident would release large quantities of gaseous fission products and activation products into the containment system. In addition, solid fission products and fuel material would be vaporized and enter the containment system, where they would quickly condense and agglomerate into suspended aerosols. The aerosols would settle to the bottom, quickly at first, then more and more slowly as their airborne concentration decreased. Comparatively large particles of fuel and solid radioisotopes that did not vaporize would initially be carried into the containment atmosphere by a pressure surge, by convection, or by a shock wave.

Containment systems are equipped with once-through high-efficiency air cleaning systems that clean effluent air as it is exhausted to the outside atmosphere. Containment systems for some types of reactors are also equipped with recirculating air cleaning systems, and one type incorporates a chemical spray system for cleanup of contained air after a postulated accident. These 3 types of post-accident air cleanup systems are shown schematically in Section II and discussed in Section VI. Testing of the efficiency of HEPA filters and charcoal beds is discussed in Section VII.

The relationships between the permissible amounts of various radionuclides that may be released from a specific reactor and the established numerical values of permissible exposures are very complex. These relationships depend on projected meteorological conditions, topography, population levels, and ecological and biological factors, but they can be calculated, using some conservatively safe estimates, and the specifications for safety features and containment of that reactor are based on these calculations (6). It is difficult to specify capacities and efficiencies of air cleaning components on the basis of estimated amounts of radioactive isotopes that will leave the reactor core and pass through the primary coolant system under normal but varying operating conditions. Calculation of the amounts of various radioisotopes that could be driven from the core by an accident and escape plateout, agglomeration and settling, and solution within the containment system and find their way to air cleaning systems or to leaks in the containment structure is the subject of continuing research, both experimental and theoretical. For practical applications the efficiencies and capacities of air cleaning systems are based on conservatively safe estimates of the maximum amounts expected to reach them under normal and accident conditions (6,7).

II. NUCLEAR REACTOR CONTAINMENT SYSTEMS

The emphasis in this section will be on the containment of power reactors. Although proper containment of every reactor is essential and required, that of power reactors is more critical to the safety of the public than that of others (except marine propulsion reactors). Were it not for the containment system, a severe postulated accident in a large high-performance power reactor could release a greater quantity of radioactive materials into the outside atmosphere than a severe accident in a research, production, or testing reactor. This is due to the large amount of fuel in the core of a power reactor and the correspondingly large amounts (discussed in Section III) of radioactive fission products, fission-product daughters, and neutron activation products that accumulate in the fuel elements and primary coolant system between refuelings. Another factor that makes containment of power reactors particularly important is the desirability of siting them close to metropolitan and industrial load centers, in order to avoid prohibitive power transmission costs.

Containment systems for power reactors are designed to withstand the pressures and temperatures resulting from design-basis accidents, which are very severe postulated combinations of malfunctions and component failures in the reactor core and primary coolant systems. In the following subsections, the most severe design-basis accident for each type of power reactor will be described briefly in conjunction with the containment for that type.

An integral part of the containment system of a water-cooled power reactor is some means of rapidly cooling the atmosphere within the containment after an accident, in order to minimize the peak post-accident pressure and to reduce the pressure to atmospheric as rapidly as possible. In pressurized-water reactors a spray system is provided for this purpose. In most boiling-water reactors a combination spray system and pool is provided. Provisions for reducing the post-accident concentration of airborne fission products within the containment atmosphere of water-cooled reactors are made as additions to these containment atmosphere cooling systems. Therefore, these systems are discussed briefly in Section II.B. (Post-accident air cleanup systems are described in Section VI.)

Gas-cooled reactors are equipped with emergency supplies of coolant gas which can be fed into the primary cooling circuit to replace coolant gas lost through leaks in the event of an accident. The coolant (usually molten sodium) in liquid-metal-cooled reactors is at ambient pressure or low pressure during normal operations, and these reactors may need no supplementary coolant if sodium flow and level are maintained.

Since containment systems for power reactors are designed to withstand major accidents, their ability to maintain their structural integrity during normal operations and minor accidents is not in question, but separate air cleaning systems are usually provided for operational releases and minor accidents, as discussed in Section V.

The containment of nuclear submarine and nuclear warship reactors are not discussed in this chapter. However, that of the maritime reactor on the N.S. SAVANNAH is described briefly under pressurized-water reactors.

A. Research, Materials Testing and Isotopes Production Reactors

Each of a number of very small research reactors, from kilowatts to a few megawatts in thermal power, is shielded by a concrete tank that contains a pool of water surrounding the reactor core. For these reactors the fuel itself, the cladding of the fuel elements, and the pool are often considered sufficient to prevent release of appreciable quantities of particulate and highly water-soluble radioactive materials. Releases of iodine-131 and its compounds and of noble gases from such small reactors might endanger operating personnel and others in the immediate vicinity, but radiation monitors, other sensing devices, and evacuation alarms are provided for their safety.

Somewhat larger reactors (about 25 to 100 megawatts in thermal power) are cooled by forced circulation of water through the reactor core and a heat exchanger. The cores and primary coolant systems of these reactors are surrounded by pressure vessels, piping, and other components designed to withstand the operating pressures plus considerable safety factors. These reactors, which are used for research, production of isotopes, and irradiation testing of materials, are much larger than the very small research reactors but much smaller than present-day power reactors. Their usual containment is a more or less conventional building in which a slight negative pressure, relative to that of the outside atmosphere, is maintained by blowers that exhaust air from within the building. The input is a large volume of leakage into the building from the outside plus any gases released from the reactor primary system. The exhaust air is passed through a high-efficiency air cleaning system designed to retain radioactive materials (except noble gases) and from there to the outside atmosphere, usually through a tall stack.

B. Water-Cooled Power Reactors

Reactors designed to generate large amounts of electrical power must produce high-pressure steam for efficient use of turbines. In the United States, two classes of water-cooled reactors are widely used for this purpose: pressurized-water reactors and boiling-water reactors. In pressurized-water reactors the coolant is not allowed to boil in the reactor core. The primary coolant water circulates through the core under higher pressure than in boiling-water reactors, is heated by the fuel elements, and circulates through a heat exchanger which generates steam in a secondary coolant system. The steam drives a turbine or turbines and is condensed, and the secondary coolant water is returned to the heat exchanger (Figure 3). In boiling-water reactors the coolant water boils in the reactor core (actually it is pressurized, but not to as high a pressure as that in pressurized-water reactors). The steam thus produced drives the turbine(s) directly; a boiling-water reactor has no secondary coolant system (Figure 2).

In water-cooled power reactors the primary coolant system (reactor pressure vessel, piping, heat exchangers, etc.) is at high pressure and temperature. These are about 1000 psi and 500° F in boiling-water reactors and as high as about 2200 psi and 600° F in pressurized-water reactors. A loss-of-coolant accident is usually considered the most serious design-basis accident. In this postulated accident, a large pipe in the primary coolant system would rupture completely, the high pressure superheated coolant water would be rapidly and freely discharged from both severed ends and flash into steam in the containment shell, and the sudden depressurization of the core would cause mechanical failure of fuel-element cladding and rapid release of fission products into the containment shell. The resulting containment atmosphere is typically assumed to consist of steam, air, and fission products at about 40 to 60 psig and about 275° to 290° F.

Emergency core-cooling systems are provided in water-cooled power reactors to remove post-accident heat from the decay of fission products in the core (which continues after the reactor is shut down), in order to minimize fuel melting and cladding rupture. This minimizes the release of fission products from the core, the reaction of cladding metal with water, and the buildup of temperature and pressure within the containment shell. If the emergency core-cooling system could not maintain enough core coolant after a loss-of-coolant accident, partial meltdown of fuel in the core would release more fission products and more energy into the containment shell. Emergency core cooling systems are complex systems of storage tanks, pumps, valves, piping, and multiple controls. Positions of the core cooling nozzles are shown schematically in Figures 6 and 7. Descriptions of these systems are given in (8).

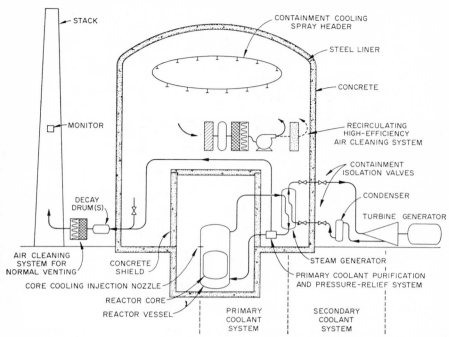

Fig. 6. Schematic of pressurized-water reactor containment, showing pressure containment shell, core and containment cooling sprays, air-cleaning systems and isolation
valves. Symbols for air cleaning components: diagonal lines for high-efficiency
moisture separators; cross-hatch for roughing filters; wavy line for HEPA filters;
and dots for impregnated charcoal. The ovoid rectangle represents a cooler. Solid
flow lines show normal air flow; broken line shows air flow after an accident.

Almost all water-cooled power reactors have provision for cooling the
containment atmosphere (separate from the emergency core-cooling system)
in the event of a loss-of-coolant accident by spraying cold water from a
spray header (Figures 6 and 7) directly into the containment shell. These
sprays are normally supplied from a tank, pool, or pressure-suppression pool,
with subsequent recycle of the water from the containment sump (or pressure
suppression pool), through coolers, and back to the spray header. In some
containment systems this cooling is vital to prevent the containment pressure
from exceeding the design limit in the event of a postulated accident.
Containment-atmosphere cooling systems are provided not only to minimize
peak post-accident pressures but also to reduce temperature and pressure
in order to reduce leakage of fission products from the containment vessel.

Cooling of the atmosphere within the containment shell is necessary during
normal operation of water-cooled power reactors to prevent excessive

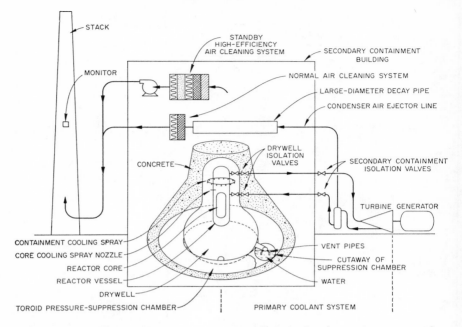

Fig. 7. Schematic of boiling-water reactor containment, showing pressure-suppression containment (drywell and torus), core and containment cooling sprays, air-cleaning systems, isolation valves and secondary containment building. Symbols for air-cleaning components: diagonal lines for high-efficiency moisture separators; cross-hatch for roughing filters; wavy line for HEPA filters; dots for impregnated charcoal.

temperatures (from heat leakage from the reactor system) that might damage electrical insulation and instruments. Operational cooling is normally provided by recirculating the containment atmosphere over water-cooled coils with a fan. Several parallel systems are usually provided which can also function to remove heat after an accident, if the fan and its motor are designed for post-accident conditions (air-steam mixtures at 40 to 60 psig and temperatures of about 290° F).

In current practice in the United States, two basic types of containment systems are applied to water-cooled power reactors. Pressure containment is used for pressurized-water reactors and a few boiling-water reactors, and pressure-suppression containment with secondary containment is used exclusively for direct-cycle boiling-water reactors. Recently, however, an ice-condenser containment system has been developed, and double containment with pumpback is under consideration for metropolitan siting.

1. Pressure Containment for Pressurized-Water Reactors

The pressure-containment shells of pressurized-water reactors are usually spherical steel pressure shells (9,10) or domed cylinders of reinforced concrete with steel liners (11,12). The containment of the 567-MWe Connecticut Yankee reactor (13), for example, is a cylinder with a domed top and a flat bottom, all of reinforced concrete. The cylindrical portion has an inside diameter of 135 ft and a height of 113 ft, and the dome has an inside radius of 67.5 ft. The total inside height is thus 180.5 ft. The cylinder is 5.5 ft thick and the dome is 4.5 ft thick (14). Both types are designed to withstand the temperature and pressure (about 40 to 60 psig) resulting from the flashing of all the primary coolant into the containment shell. The containment system of a pressurized-water reactor is shown schematically in Figure 6.

For pressurized-water reactors, a maximum acceptable design-basis-accident leakage rate (usually 0.1 wt.% of the contained volume per 24 hr) is specified for the shell at the maximum design-basis-accident temperature and pressure (15).

In a modification of the pressure-containment system, a secondary enclosure is placed entirely or partially around the containment shell. The containment shell serves the same purpose as in simple pressure containment. However, any leakage from it is to the volume between the two structures, in which fission products can be held up for radioactive decay, diluted with outside air from secondary-containment inleakage or supply fans, cleaned by a high-efficiency air cleaning system, and exhausted through a high stack. This type of containment is not yet in wide use for pressurized-water reactors. Each of one set of twin reactors (9) (1971–1972 startup) has a spherical steel pressure-containment shell, as well as a penetration room attached to the outside of the containment shell in a position to enclose many of the containment-shell penetrations which serves as a partial secondary containment structure. Another pressure containment shell is provided with pressurization of welds and penetrations to ensure low leakage (12).

Another modification of pressure containment proposed for metropolitan siting (16,17) provides double containment with pumpback. This containment system will have two domed steel cylinders, separated by a 20-in. thick annulus filled with porous concrete. The annulus will be kept at a pressure slightly lower than that of the outside air, and any leakage will be pumped back into the inner containment shell. A diagram of a double containment and its air cleaning system is contained in Section VI, with a description of the treatment of contained air after a postulated accident.

In the recently developed ice-condenser reactor containment system (18) an insulated blanket of millions of hollow cylinders of ice forms a massive heat sink around the inside of the containment shell. The ice is kept frozen by standard refrigeration equipment. In an accident, insulated panels would automatically spring open to provide a path for steam to reach the ice. In one test (18) the steam pressure was reduced to a safe level in 30 sec, and not over 2 psig pressure was built up in the containment shell. In the ice-condenser containment design, the design pressure requirement is reduced from about 45 psig to 10 psig, and the containment-shell volume is reduced by half. This type of containment system is planned for the Cook reactors (19) (1972 startup) and proposed for the Sequoyah reactors (20) (1974 startup).

The reactor that powers the main drive turbines of the N.S. SAVANNAH (21) deserves special mention here because, although it is relatively small (about 75 MW thermal power), it is a power-producing pressurized-water reactor and its primary coolant system operates in the same pressure and temperature ranges used in stationary pressurized-water power reactor plants. The reactor and primary system are surrounded by a horizontal cylinder with domed ends, with a smaller vertical domed cylinder on top. This containment vessel is designed to contain an internal pressure of 173 psig, the calculated maximum pressure in the event of rupture of the primary system while the reactor is operating at full power. A secondary containment compartment maintained at a slight negative pressure by controlled inleakage and exhausted through high-efficiency filters and charcoal adsorbers completely surrounds the pressure containment.

2. Pressure-Suppression Containment with Secondary Containment for Boiling Water Reactors

Pressure-suppression containment shells for boiling-water reactors are divided into two compartments (22–25). The reactor and most of the primary coolant system are enclosed in a pressure shell shaped like a light bulb with the base upward (see Figure 7). This shell, conventionally referred to as the drywell, is designed to withstand a pressure of about 60 psig from flashing of the primary coolant. The drywell is surrounded by a toroidal (doughnut-shaped) suppression chamber with its horizontal centerline a little below the bottom of the drywell. The containment system of a typical boiling-water reactor is shown schematically in Figure 7.

The drywell of the 515-MWe Oyster Creek Reactor (23) has a "bulb" (major) diameter of 70 ft, a "base" (minor) diameter of 32 ft, and an inside height of 120 ft. The toroidal suppression chamber has a cross-sectional (minor) diameter of 29 ft and an average major diameter (distance between centers of the two opposite cross sections on a major diameter) of 101 ft (14).

The gross major diameter is thus about 130 ft, excluding the surrounding concrete.

The suppression chamber, often called the wetwell or the torus, is about half full of water, is subject to lower postulated accident pressures than the drywell, and is designed to withstand a pressure of about 35 psig. The drywell is connected to the suppression chamber by large vent pipes that discharge 3 or 4 ft under the surface of the water. In the event of a rupture of the primary system in the drywell, the coolant water would flash into steam, and the pressure would force it through the vent pipes and under water in the suppression chamber, where it would condense. The water in the suppression chamber would thus serve as a heat sink for the almost immediate absorption of most of the energy in the released primary coolant. Noncondensable gases would be vented back to the drywell to relieve pressure in the suppression chamber.

Pressure-suppression systems have a refueling building surrounding the containment shell, which acts as a secondary containment structure. Provision for removal of fission products from the atmosphere within this secondary containment building is essential. In these direct-cycle systems, the primary steam line to the turbine and the condensate return line extend the primary coolant system outside the drywell, and an integral part of the containment concept is quick-acting isolation valves on these lines. If a pipe ruptured in the part of the primary coolant system outside the containment shell, some primary coolant would leak into the secondary containment building. The isolation valves would close rapidly; if the coolant contained fission products at the time of rupture, however, some would be discharged into the atmosphere within the secondary containment building. Once-through filter-adsorber air cleaning systems are provided to clean the air before it is exhausted from the secondary containment building.

For typical boiling-water reactors with pressure-suppression containment, the maximum acceptable design-basis-accident leakage rate for the drywell and the pressure-suppression chamber is usually 0.5 wt.% of the contained volume per 24 hr (15). This is 5 times the corresponding leakage rate for the pressure containment of pressurized-water reactors, but the leakage enters the secondary containment structure, from which it is exhausted through a high-efficiency air-cleaning system as mentioned above.

C. Gas-Cooled Power Reactors

There are a number of types of gas-cooled power reactors in various stages of development. The greatest emphasis, however, has been placed on two types: graphite-moderated reactors cooled with carbon dioxide (CO_2), and high-temperature reactors cooled with helium. A large fraction of the electrical power generated by nuclear reactors is produced by CO_2-cooled, graphite-moderated reactors. In fact, in 1967 about two-thirds of the world's nuclear power was generated in the United Kingdom and France by reactors of this type (26). Helium-cooled reactors are in the prototype stage, but they show considerable promise for use as thermal (moderated) reactors or as fast (unmoderated) reactors.

1. Steel or Prestressed Concrete Pressure Vessels for CO_2-Cooled Graphite-Moderated Reactors

For their first stage of large-scale nuclear power production, both the United Kingdom and France chose CO_2-cooled reactors fueled with metallic natural uranium and moderated with large quantities of graphite in the core. In the context of reactor technology, natural uranium is uranium that has not been enriched in the fissionable isotope uranium-235; it contains only a little over 0.7% ^{235}U. For efficient use of this fuel, the fuel elements are clad with one of several magnesium alloys of very high magnesium content (about 98.3 to 99.5%, depending on the alloy). Magnesium absorbs less thermal neutrons than other widely used cladding materials and is compatible with metallic uranium. In English-speaking countries, these reactors have become known as Magnox reactors, from the trade name for the principal magnesium alloy used as cladding in the United Kingdom. The British reactors of this type are also called Stage 1 power reactors.

The earliest Magnox reactors were low in power (40 to 60 MWe) and also low in power density (thermal power per unit volume of core). During the past decade they have been rapidly scaled up in power; some of the newest are in the 600-MWe range. In most of the British Magnox reactors and a few others, the core is contained in a steel pressure vessel through which the CO_2 is circulated. A concrete biological shield surrounds the steel vessel, and the space between the two is air cooled. The outer containment is a conventional building (a steel shell was used in one case for experimental reasons (27)). This type of containment is shown schematically in Figure 8. The spherical pressure vessel of each of the twin 250-MWe Hinkley Point A reactors is 74 ft in diameter and 3 in. thick (28).

Unlike the primary coolant water in water-cooled reactors, which could flash into steam in a postulated accident, the gas in CO_2-cooled reactors cannot cause a rapid increase in containment pressure by changing from a liquid to a gas. Provision must however be made for preventing the core from overheating and melting the cladding in the event of a possible breach in the pressurized CO_2 circuit. Also, the large quantities of graphite moderator must be protected from combustion as a result of reaction with the shield cooling air. A large emergency supply of CO_2 is kept on standby to replace coolant gas losses in case of accidental depressurization.

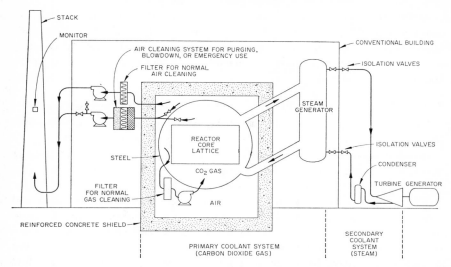

Fig. 8. Schematic of Magnox gas-cooled reactor (coolant, carbon dioxide), showing steel pressure vessel, shield cooling air, concrete biological shield, air-cleaning systems, and isolation valves. Symbols for air-cleaning components: cross-hatch for roughing filters; wavy line for HEPA filters; dots for impregnated charcoal.

In most French reactors of the Magnox type, the core is contained in a pressure vessel of prestressed concrete rather than steel. Prestressed concrete is held in compression under normal conditions by steel cables which are anchored at the ends and are under tension. Each pressure vessel is prestressed by a number of sets of cables which are oriented according to the configuration of the vessel. Each set contains many cables to minimize both the peak accident tension in any one cable and the consequences of breakage of one or more cables. The design philosophy of prestressed concrete vessels is that the concrete would develop leaks under excessive pressure, the leaks would relieve the pressure, and the tension of the cables would reseal the leaks to a great extent, restoring a considerable degree of leak-tightness. The prestressed concrete vessel of the EDF 3 reactor in Chinon, France, which is cooled with an organic liquid, is cylindrical inside and cubic outside, with an inside diameter of 62 ft, an inside height of 69 ft, and a thickness of 14 ft (14).

The largest of the British Magnox reactors use prestressed concrete pressure vessels with steel liners. The liner is insulated and cooled by water which is circulated through pipes in the concrete (29). The spherical steel liner of the Wylfa reactor's vessel is 96 ft in diameter and $\frac{3}{4}$ in. thick. The prestressed concrete vessel is 11 ft thick (14).

Advanced CO_2-cooled, graphite-moderated reactors in the United Kingdom use uranium dioxide fuel artificially enriched in fissionable uranium-235, stainless steel fuel-element cladding, and prestressed-concrete pressure vessels. These reactors, most of which are still under construction, are in the 600-MWe range (30,31).

2. Steel-Lined Prestressed Concrete Pressure Vessels for Helium-Cooled Reactors

Large high-temperature helium-cooled power reactors as presently designed (and under construction in one case (32)) will use prestressed-concrete pressure vessels, with water-cooled steel liners, as their inner containment vessels. The outer containment will probably be a conventional building equipped with filters and a stack, as shown in Figure 9 (26).

The fuel elements will be hexagonal blocks of graphite. Each block will contain a number of longitudinal holes in a circularly symmetrical pattern. Small-diameter holes will hold sticks of bonded fuel particles; larger holes will be channels for the helium coolant. The graphite blocks will serve as the structural members of the core and provide heat transfer between the fuel and the coolant (26). The fuel sticks will not be clad. Instead, the fuel particles will be clad individually before they are bonded. Each particle is coated with a pyrolytically deposited inner layer of porous carbon, which provides volume for the accumulation of gaseous fission products. An outer layer of high-density carbon, also pyrolytically deposited, is the basic cladding of the fuel particle. This layer is protected from high-velocity fission fragments by the porous inner layer (26).

Existing high-temperature helium-cooled reactors have double or triple containments. The Peach Bottom and Dragon reactors have steel primary and secondary vessels, between which an inert atmosphere of nitrogen can be maintained. Nitrogen is used in the Peach Bottom reactor; air has been used in the Dragon reactor. The Julich AVR has three steel vessels. The space between the inner two is kept filled with nitrogen, and the outer vessel is filled with air (33). These three reactors are experimental, developmental, and prototype models.

As described above, a single pressure vessel, of prestressed concrete, is expected to be used on large high-temperature helium-cooled reactors. However, the pressure vessel will contain not only the reactor core and some of the circulating primary coolant; it will contain the whole primary coolant circuit, including the steam generators and the helium circulators. In Figure 9, two steam generators are represented by the two sets of unshaded vertical cylinders inside the pressure vessel, and a circulator is represented by the cylinder protruding into the bottom at the center. Actually there are several steam generators and several helium circulators, symmetrically arranged.

HTGR CONTAINMENT SYSTEM FEATURES

Retention Devices	Function
1. Coated Particles	Retains
2. Fuel Element	Delays
3. Helium Circuit	Delays and Dilutes
4. PCRV	Contains
5. Building	Dilutes
6. Filter	Retains or Delays
7. Stack or Vent	Elevates and Dilutes
8. Helium Cloud	Elevates and Dilutes

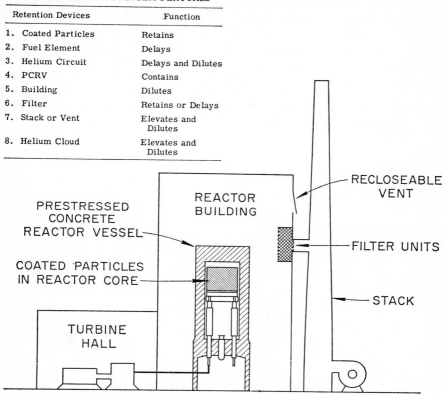

GCR Containment System

Fig. 9. Schematic of high-temperature gas-cooled reactor (coolant, helium), showing prestressed concrete reactor vessel, filter units, and secondary containment building. Air-cleaning system components, not specified in the figure, will include roughing and HEPA filters and impregnated charcoal (26).

D. Liquid-Metal-Cooled Fast Breeder Reactors

In the design of liquid-metal-cooled fast breeder reactors, the current emphasis is on mixed uranium and plutonium oxide or carbide fuels, sodium coolant with argon blanket gas (both with cleanup systems), a nitrogen-filled containment system surrounding the coolant-blanket gas system to prevent combustion of sodium in case the coolant system is ruptured, and an outer containment structure equipped with an air cleaning system to control sodium oxide aerosols, fission products, and plutonium in case of postulated accidents.   The containment system of a sodium-cooled fast breeder reactor is shown schematically in Figure 10.

In liquid-metal-cooled reactors, a high coolant temperature (500 to 650° C) is possible without pressurization of the primary coolant.   This is a considerable safety advantage in cases of rupture of the coolant system.   Sodium melts at 97.8° C and is therefore easy to keep liquid in a power-plant system. It boils at 882° C and thus permits adequately high coolant temperatures with the system at atmospheric pressure.

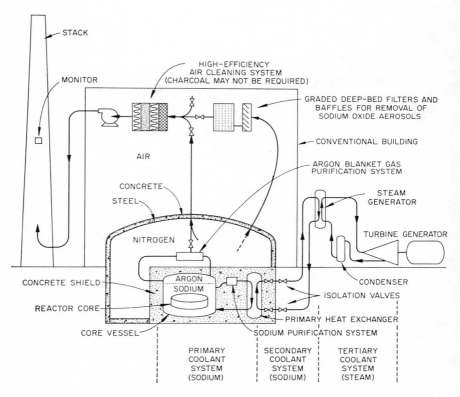

Fig. 10. Schematic of liquid-metal-cooled reactor containment, showing purification systems, for primary coolant and argon blanket gas, and air-cleaning system, for nitrogen containment gas, with baffles and deep-bed filters for removal of sodium oxide aerosols if necessary. Symbols for air-cleaning components: partial diagonal lines for baffles; columns of dashes for graded deep-bed filters; cross-hatch for roughing filters; wavy line for HEPA filters; dots for impregnated charcoal. Broken flow line shows air flow after an accident.

Sodium has certain disadvantages, however. It reacts chemically with oxygen and water and must therefore be isolated from them. Further, exposure of sodium to neutrons produces ^{24}Na, a strong gamma emitter with a half-life of about 15 hr. Therefore in current designs the steam boiler has been separated from the primary sodium coolant by a secondary sodium coolant, which does not become radioactive. The vault necessary to contain the radioactive primary sodium provides the shielding required and also acts as a container for the inert blanket gas (usually argon) that surrounds the primary system.

In these reactors, the fuels will be operated to high burnup and thus will accumulate large fission-product inventories. Expansion of high-burnup fuels due to solid fission products must be accommodated by low effective fuel density (compacted powders, pressed and sintered pellets, or axial holes) or by annular expansion volumes between fuel and cladding. The annulus can be filled with gas or sodium.

The buildup of pressure from volatile fission products could be contained by thick cladding, but this is expensive in terms of breeding ratio. The cladding can be extended to provide a plenum (reservoir) above or below the core which reduces the pressure buildup in the fuel element by increasing the volume within the cladding.

Gaseous fission-product pressure could become quite high in vented-to-plenum fuel elements because, although the plenum would be outside the core, it would be an integral part of the cladding. This can be avoided by venting the fuel element into the primary sodium coolant. Current development work on venting devices emphasizes holdup within the fuel element long enough for short-lived fission-product gases to decay and for their decay products to deposit within the cladding; thus the long-lived fission-product gases would be selectively released. Venting should lead to less frequent failure of fuel elements and thus reduce contamination of the coolant through leaching of the fuel by the coolant.

The advantages of venting fuel to the sodium coolant must be weighed against the possibilities of increased shielding requirements, additional decontamination problems in normal maintenance, and increased hazard to personnel from primary-coolant-system leaks involving few or no damaged fuel elements. In the case of vented sodium-bonded carbide or alloy fuels, the possibility of loss of bond before bulk coolant boiling occurs must also be considered.

In the event of rupture of the primary coolant system, sodium containing radioactive ^{24}Na, fission products, and possibly fuel materials would be released into the nitrogen-filled containment structure enclosing it. If the containment were breached by the pressure of sodium-vapor expansion, a pressure surge, or a postulated nuclear excursion, these materials would escape into the air atmosphere within the secondary containment system, the sodium would burn, and aerosols of sodium oxide and other materials would be produced. A large fraction of the fission products and fuel materials (but not noble gases) might be occluded by, attached to, or sorbed by the sodium oxide aerosols, most of which would agglomerate rapidly into particles that could be removed by filtration. The filtration problems created by fine, uncoagulated aerosol particles might be aggravated by the presence of radioactive ^{24}Na. On the other hand, most of the iodine would probably be in the form of NaI or otherwise associated with sodium. This might reduce or possibly eliminate the necessity for removal of gaseous iodine and methyl iodide by charcoal beds. To prevent excessive loading of the high-efficiency filters, high concentrations of sodium oxide aerosols must first be reduced by baffles and/or graded deep-bed filters of stainless steel or other fibers (34).

III. IMPORTANT AIRBORNE RADIOACTIVE MATERIALS IN CONTAINMENT SYSTEMS

As the fuel in a nuclear reactor fissions ("burns up"), fission products accumulate within the sealed cladding of each fuel element. Accumulation of solid and of a part of the liquid and gaseous fission products causes the fuel to swell gradually. The expanded fuel and the gaseous fission products exert increasing pressure on the cladding. Modern cladding alloys and designs produce fuel elements of very high structural integrity, but some fuel elements develop leaks as a result of internal pressures, thermal cycling, corrosion, etc. In addition to radioactive fission products, radioactive isotopes are produced by neutron activation (reaction with neutrons) of some of the elements in the coolant (and in core structural materials). Radioactive fuel might also become airborne in a postulated major accident.

A. Radioactive Fission Products in Reactor Fuels

As discussed in Section I, it is very difficult to estimate how much of each radioisotope in a reactor core could be released from the core, by leaking fuel elements or by an accident, and make its way to an air-cleaning system or to a leak in the outer containment. In operating nuclear reactor plants, normal releases are measured by continuous monitoring. Once established for a specific type of fuel element in a particular reactor, normal releases can be predicted with reasonable accuracy for fuel elements of the same type, provided they are used in a reactor of the same type and under very similar operating conditions. Calculations of releases from the core under postulated accident conditions can only be semiquantitative and must be conservatively high.

On the other hand, the quantities of fission products in a given reactor core at a specified time can be accurately estimated from the history of that core (35,36). Iodine-131 and krypton-85 are the most important radioactive fission products in the control and containment of airborne radioactive materials in nuclear reactor plants. The other isotopes listed in Table I (argon-41, xenon-133, and tritium) also require careful monitoring and control. In addition to these, it is possible that the fission products cesium, tellurium, strontium, barium, and ruthenium could present significant hazards under reactor accident conditions.

Over 200 isotopes are formed in the fission process. The more important of these are listed in Table II (35,36), approximately in order of the volatility of the elements but with some changes dictated by observed behavior in fission-product release experiments. The table gives the quantities produced (radioactivity in kilocuries and weight in grams) per megawatt of thermal power in the core after one year of irradiation (equivalent to one year of continuous operation, or 8760 megawatt-hours of thermal power). Boiling points, probable chemical forms under accident conditions, and radiological hazards are also listed.

Since there are so many radioactive (and stable) isotopes in the fission spectrum, it is desirable to simplify the analysis, by reducing the number of species under consideration. One such scheme (35) is shown in Table III. A study of Table II shows that krypton, xenon, and iodine are very important, because they are volatile and are present in large quantities. In addition, iodine concentrates in the thyroid and, consequently, is biologically potent. Radioactive cesium, even though its yield is low, is moderately volatile and, when ingested, constitutes a hazard to the whole body. Tellurium has very short-lived high-yield isotopes. It also is moderately volatile and contributes a health hazard through the ^{132}I daughter product of ^{132}Te. Strontium has a relatively high radioactivity and is a long-lived hazard to bone and lung, even though it is relatively nonvolatile. Barium is of high yield and slight volatility and hazardous to bone and lung tissue. Ruthenium is highly volatile under strongly oxidizing conditions. Other isotopes in Table II are excluded from Table III because of low volatility, low fission yield, relatively low toxicity, or very short half-life.

The time in core and the power level of a reactor are major factors in determining the inventory of fission products. The short-lived isotopes reach their saturation concentration soon after reactor startup, while the long-lived activities, those with half-lives in the range months to years, continue to accumulate during an average reactor core life. After extended irradiation, about 50% of the total activity is always associated with the refractory elements (the rare earths, zirconium, and niobium). The halogens, rare gases, alkaline-earth metals (barium and strontium), and

TABLE II

Yields and Characteristics of Important Fission Product Isotopes Arranged Approximately in the Order of the Volatility of the Elements[a] (35 and 36)

Isotope	Half-life	Activity [Kilocuries/Mw(thermal)] after 1 yr of irradiation[b]		Total weight [g/Mw(thermal)] after 1 yr of irradiation		Boiling point[c] (°C)	Probable release and transport form[d]	Hazard
		At shutdown	1 day after shutdown	Isotope	Element			
		High Volatility (Group I)						
Kr	Stable	—	—	4.9		−153	Elemental gas	External radiation, slight health hazard
-83m	114m	3	0					
-85	10.27y	0.1	0.1	0.3				
-85m	4.4h	8	0.2					
-87	78m	15	0					
-88	2.8h	23	0.1					
-89	3m	31	0					
-90	33s	38	0		5.2			
Xe	Stable	—	—	55.4		−108	Elemental gas	External radiation, slight health hazard
-131m	12d	0.3	0.3	0.003				
-133m	2.3d	1	0.7	0.003				
-133	5.27d	54	47	0.3				
-135m	15.6m	16	0					
-135	9.2h	34	14	0.01				
-137	3.9m	48	0					
-138	17m	53	0					
-139	41s	61	0		55.7			
Br	Stable	—	—	0.03		59	Elemental Br$_2$ or HBr	External whole body radiation, moderate health hazard
-83	2.3h	3	0					
-84	32m	6	0					
-85	3m	8	0					
-87	56s	15	0	0.3				

Isotope	Half-life				Total	Chemical form in reactor[d]	Biological hazard
I	Stable	—	—	0.45	183	Elemental gas I_2 or HI; variable gaseous organic compounds; adsorbed forms on small particles	External radiation, internal irradiation of thyroid, high radiotoxicity
-129	1.7×10^7 y	10^{-6}	10^{-6}	7.92			
-131	8d	25	23	0.20			
-132[e]	2.3h	38	0	0.004			
-133	21h	55	26	0.05			
-134	52m	63	0	0.002			
-135	6.7h	55	4.4	0.01			
-136	86s	53	0	0.000			
					2.6		
Cs	Stable	—	—	22.7	685	Elemental Cs converting to CsOH	Internal hazard to whole body
-134	2y	0.029	0.29	0.03			
-136	13d	0.147	0.147	0.002			
-137	26.6y	1.28	1.3	12.2			
					35.0		
Intermediate Volatility (Group II)							
Te	Stable	—	—	5.1	987	Elemental Te above 1000°C, converting to TeO_2	External radiation, moderate health hazard
-125m	58d	0.005	—	0.0003			
-127m	105d	0.414	0.5	0.04			^{132}Te contributes a health hazard from ^{132}I daughter
-127	9.4h	2.1	0.5	0.2			
-129m	34d	2.9	2.3	0.007			
-131m	30h	3.9	2.2	0.1			
-131	25m	26	30.1	0.1			
-132	77h	37.3	0				
-133m	52m	54	0				
-134	44m	57	0				
-135	2m	36					
					5.4		

[a] This table is an expansion of a similar table published by J. R. Beattie in "An Assessment of Environmental Hazard from Fission Product Releases," AHSB(S)R-64 (May 16, 1961).

[b] Assumed thermal neutron flux, 5×10^{12} neutrons/cm².

[c] See J. O'M. Bockris, J. L. White, and J. O. Mackenzie (eds.), "Physicochemical Measurements at High Temperatures," Appendix VI (By J. L. Margrave), Academic Press, New York, 1959; also see A. N. Nesmeyanov, "Vapor Pressure of the Chemical Elements" (R. Gary, ed.), Elsevier, New York, 1963.

[d] Chemical states of released fission product elements are affected by temperature and by composition of the ambient atmosphere during and after release. Those shown in the table are selected to represent most probable chemical states during typical accident conditions.

[e] 38.0 kilocuries of I-132 are produced by decay of Te-132 in the reactor. Te-132 released in an accident will produce more I-132.

TABLE II continued

Isotope	Half-life	Activity [Kilocuries/Mw(thermal)] after 1 yr of irradiation[b]		Total weight [g/Mw(thermal)] after 1 yr of irradiation		Boiling point[c] (°C)	Probable release and transport form[d]	Hazard
		At shutdown	1 day after shutdown	Isotope	Element			
			Volatile Under Highly Oxidizing Conditions (Group III)					
Ru	Stable	—	—	17.4		4230	Volatile oxide, RuO_4	Internal hazard to kidney and GI tract
-103	41d	25.7	25.7	0.8	18.7			
-106	1y	1.54	1.54	0.5				
Tc-99	2.12×10^5 y	10^{-2}	10^{-2}	9.3	9.3	(4600)	Volatile oxide, Tc_2O_7	Internal hazard to GI tract and lung
-99m	6.04h	5.5	0.1	0.001				
Mo	Stable	—	—	34.3	34.4	(4800)	Volatile oxide, MoO_3	Internal hazard to GI tract and lung
-99	67h	51.5	40	0.1				
			Low Volatility (Group IV)					
Sr	Stable	—	—	6.5		1366	Elemental Sr converting to SrO	Internal hazard to bone and lung
-89	54d	39	39	1.5				
-90	28y	1.2	1.2	8.7				
-91	9.7h	51	9.7	0.0	14.6			
Ba	Stable	—	—	14.6		1635	Elemental Ba converting to BaO	Internal hazard to bone
-140	12.8d	53	48	0.7	15.3			
Sb-125	2.7y	0.04	0.041	0.10	0.10	1640	Elemental Sb or Sb_2O_3	Internal hazard to GI tract, lung, total body, and bone

Refractory (Group V)

Isotope	Half-life					m.p.	Oxide form	Internal hazard
Sm	Stable	—	—	5.2		1602	Elemental Sm converting to Sm_2O_3	Internal hazard to bone, lung, and GI tract
-151	93y	0.01	0.01	0.42				
-153	47h	1.4	1.0	0.003				
-156	~10h	0.1	0.02	0.00	5.6			
Pm-147	2.6y	4.8	4.8	5.0	5.0	(2700)	Elemental Pm converting to Pm_2O_3	Internal hazard to GI tract, bone, and lung
-149	54h	11.8	8.7	0.02				
Pr	Stable	—	—	11.4		3020	Elemental Pr converting to Pr_2O_3	Internal hazard to GI tract
-143	13.7d	52	45	0.8				
-145	6h	36	2.34	0.01	12.2			
Y	Stable	—	—	4.9		2783	Elemental Y converting to Y_2O_3	Internal hazard to GI tract
-90	64.5h	1.2	0.9	0.002				
-91	58d	53	52.4	2.2				
-92	3.6h	52.6	0.5	0.006	7.1			
Nd	Stable	—	—	39.9		3090	Elemental Nd converting to Nd_2O_3	Internal hazard to GI tract, liver, and lung
-147	11.3d	23.2	20	0.3	40.2			
La	Stable	—	—	12.9		3370	Elemental La converting to LaO_2	Internal hazard to GI tract
-140[f]	40h	54	36	0.1	13.0			
Ce	Stable	—	—	23.9		3470	Elemental Ce converting to CeO_2	Internal hazard to liver and lung
-141	32d	55.3	55.3	1.8	35.2			
-143	33h	54	33	0.053	36.8			
-144	290d	30.0	30.0	9.5				
Zr	Stable	—	—	43.5		4325	Elemental Zr converting to ZrO_2	Internal hazard to GI tract, total body, and lung
-95	63d	53	53	2.4	45.9			
Nb	Stable	—	—	0.0		4930	Elemental Nb converting to Nb_2O_5	Internal hazard to GI tract, bone, lung, total body
-95m	90h	0.5	0.5	0.001				
-95	35d	53	53	1.3	1.3			

f See Ba-140.

TABLE III

The Most Hazardous Fission Product Activities
in Order of Decreasing Volatility (35)

| Element | Activity (kilocuries/Mw (thermal)) After 1 yr of irradiation | | Weight (g) |
	At shutdown	1 day after shutdown	
Xe, Kr	385	62	61
Iodine	289	53	2.6
Te, Cs	220	37	40
Ru	27	27	19
Sr, Ba	144	98	31

tellurium are each present initially to the extent of 10 % to 15 % of the total activity. Tellurium falls below 8 % after 24 hours or less, and iodine drops below 6 % in not more than 10 days. The barium-strontium group is the only one of the four mentioned above that gains in abundance after 10 days. The refractory group continues to increase to 70 % to 80 % of the total after this short period. For periods longer than 50 days, ^{106}Ru, ^{137}Cs, and ^{90}Sr become the most significant radioactivities in addition to the refractory group.

B. Airborne Fission Products During Normal Operations

For practical purposes, the amounts of most gaseous, volatile, and liquid fission products that escape from fuel elements into the primary coolant during normal operations are functions of the number of leaking fuel elements and the effective sizes of the leaks in their claddings. (Tritium is an exception to this, as discussed below.) Even in leaking fuel elements, most of the solid fission products are leached only slowly from the solid fuel, and particulate radioactive materials in the coolant will be largely activated corrosion products, with some precipitated or occluded fission products. Generally, the amount of solid material that is sufficiently loosened from the fuel to be swept away mechanically by the coolant is negligible.

Most volatile, liquid, and gaseous radioactive fission products released from the core during normal operations are produced in such small quantities or have such short half-lives that their radioactivities are at or near the limits of detection, or are retained in the coolant in water-cooled or liquid-metal-cooled reactors. Of those that can become airborne and are radioactive enough and long-lived enough to present potential hazards, krypton-85 and iodine-131 far outweigh the others in importance. Xenon-133 and tritium (hydrogen-3) merit brief attention.

Krypton-85 and xenon-133 are chemically inert. Their only short-term hazard is a possible external dose to the whole body of a person immersed in the effluent air containing a large quantity of one or both of these isotopes released suddenly, by an accident, under unfavorable meteorological conditions. Xenon-133 is not of long-term significance; its half-life is 5.3 days. Krypton-85, however, has a half-life of about 10.3 years, and its long-term accumulation in the earth's atmosphere is expected to cause problems in the future unless it can be reduced.

Although tritium is a gas, the fission-product tritium released from fuel elements is reacted to form tritiated water (HTO), and the amount that becomes airborne during normal operations is not significant. There is evidence of tritium diffusion through intact stainless steel claddings (37,38). Tritium is also produced in coolant water by neutron activation (Section III-C).

Iodine-131 has a very low maximum permissible concentration, 3×10^{-9} microcuries per cubic centimeter of air for continuous ingestion, because it concentrates in the thyroid gland. It does not represent a cumulative long-term hazard in the atmosphere (half-life, about 8 days), but it is considered the most critical radioactive material in the treatment of airborne wastes from reactors, because of its effect on the thyroid.

C. Airborne Activation Products During Normal Operations

The principal gaseous activation products to be considered are argon-41, nitrogen-13, and tritium (hydrogen-3). Nitrogen-16 and oxygen-19 are produced in some quantity. These gases have half-lives of only seconds and decay to insignificant quantities during their passage through ducts, air-cleaning systems, delay lines, and stacks. Most of these activation gases are produced by neutron activation of elements in the liquid or gaseous coolant proper. However, argon-41 is produced by activation of argon-40 in shield cooling air as it flows between the steel reactor vessel and the concrete biological shield of Magnox reactors, as well as by activation of argon-40 in the CO_2 coolant.

Since argon is chemically and biologically inert, argon-41 does not present an ingestion hazard. The only significant effect would be gamma and some beta irradiation of persons exposed to the effluent plume from the reactor system before it had become widely dispersed. It will undoubtedly be produced by activation of the argon blanket gas proposed for liquid-metal-cooled fast breeder reactors. The half-life of argon-41 is long enough (about 110 min) to necessitate careful regard for its short-term radiation, but far too short to produce long-term accumulation in the atmosphere.

In addition to its formation as a fission product, tritium is produced in the primary coolant water of water-cooled reactors by activation of naturally occurring deuterium (hydrogen-2) and by activation of boron or lithium when they are used as additives.   A very small portion of the tritium becomes airborne.   Roughly 99 % of the tritium is released as liquid tritiated water, and even the effluent water has contained less than 1 % of the maximum permissible concentration.   In Magnox gas-cooled reactors, tritium is pro-duced from lithium present as an impurity in the graphite moderator; almost all of this tritium is condensed and discharged as a liquid.   Although tritium has a fairly long half-life (12.6 years), it rapidly reaches a very low equilibrium concentration when distributed through all the body tissues, and is therefore not considered a significant hazard.   As a matter of fact, the total release of tritium from fission reactors, including tritium dis-charged as a liquid, is considered (37) to pose no severe worldwide population exposure problems, although it does pose handling problems in fuel reprocess-ing plants and may lead to local environmental contamination problems in large fuel reprocessing plants.

Nitrogen-13 is produced in considerable amounts by activation in boiling-water reactors.   Its half-life is quite short, about 10 minutes, and it assumes relative importance only in the absence of releases of noble gases from leaking fuel elements.   Nitrogen-13 is not listed in the tables (4,5) on which Table I is based.

D. Airborne Radioactive Materials Under Postulated Accident Conditions

In the event of a postulated accident in any power reactor, the primary coolant system could be ruptured, as described in Sections I and II.   Fuel-element claddings could be ruptured or melted, and the fuel could overheat, melt down, and vaporize.   Some quantity of every radioactive fission product, neutron activation product, and fuel isotope could be released into the inner safety containment vessel.   The vaporized fuel compounds (largely oxides of uranium, plutonium, or thorium or combinations of these) and vaporized solid fission products and activation products would condense into tiny aerosol particles, agglomerate into larger particles, and begin to settle out, accompanied by compounds from cladding and core structural materials (and sodium in sodium-cooled reactors).

If the inner safety containment vessel were breached, a part of each of these materials could be released into the air atmosphere within the outer containment, from which they would be removed by air-cleanup systems. These are special air cleaning systems (Section VI) provided in all power reactors for use in case of a severe postulated accident.

Under these conditions, the radioactive materials that became airborne within the outer containment would include the fission products and activa-tion products discussed in Sections III-B and III-C (in much larger quan-tities than under normal operating conditions), various quantities of the isotopes listed in Table II, and some fraction of all fuel isotopes used or produced in the reactor.   Of the fission products in Table II, xenon, cesium, tellurium, strontium, barium, and possibly ruthenium would present significant hazards, as discussed in Section III-A.   However, iodine-131 and krypton-85 would still be most important.   Sodium-24, a product of neutron activation of the sodium coolant, would present a short-term hazard in sodium-cooled reactors.

Fuels include heavy fissile and fertile isotopes, primarily some of the isotopes of uranium, plutonium, and thorium.   Although they are all hazardous, plutonium-239 is considered far more dangerous than the others, because of the combination of its high radiological toxicity and its presence in large quantities.   In low-enrichment reactors, neutron capture by ^{238}U continuously produces 2.3-day ^{239}Np.   In fast breeder reactors, formation and decay of ^{239}Np will produce large amounts of ^{239}Pu, in addition to the ^{239}Pu in the fuel at startup.   Neptunium and plutonium both form stable oxides (NpO_2 and PuO_2) which are comparable in volatility to UO_2, and a large fraction would be expected to remain in the reactor core or the primary coolant under most accident conditions.   The fuel materials released into the outer containment would be in the form of solid particles, or "particulates," as discussed in Section IV-A.

IV. HIGH-EFFICIENCY AIR CLEANING SYSTEMS FOR NUCLEAR REACTORS

Air cleaning systems provided for cleanup of reactor containment atmospheres after postulated accidents are designed to remove and retain high percentages of the airborne radioactive materials, in order to assist the low-leak-rate containment systems in keeping the release of radioactive materials below the allowable limits for the release of airborne radioactive materials into the environment after a postulated major accident.   In order to do this, these air cleaning systems must maintain high efficiency, reliability, and structural integrity (including leaktightness) after a postulated accident. Systems for normal cleaning of reactor off-gases are designed to delay and control the release of radioactive noble gases and to remove and retain all but a very small fraction of other radioactive materials in the off-gases.

To serve these purposes effectively, reactor air cleaning systems have unique design requirements.   They are not required to have the ultra-ultrahigh collection efficiency of some filters for microbiological use (collection of every hazardous microorganism in the filtered air in some cases), nor are they required to have the tremendous flow capacities of many large conventional air-cleaning systems.   What makes reactor air cleaning systems unique is the combination of requirements that they must meet.   They must have ultrahigh efficiencies, far higher than those of ordinary industrial systems, and must not fail.   And yet these systems must have moderately high flow capacities (despite low flow through the filter media), low pressure drop, and acceptable dust loadings, without being prohibitively bulky or expensive.

Such systems have been made economically practical during the last 10 years by the development of two key components: high-efficiency particulate air filters (HEPA filters) for the removal of submicron particles, and activated charcoal specially impregnated for the removal of methyl iodide as well as molecular (elemental) iodine.

A. Classification of Airborne Radioactive Materials for Air Cleaning Purposes

For practical purposes in the design of air-cleaning systems, airborne radioactive materials are usually divided into four groups: comparatively large entrained particles, tiny aerosol particles, iodine and other chemically reactive gases, and noble gases.   (Aerosols and larger particles are often considered together, and referred to as particulate materials or "particulates.")   A fifth component for which trapping must be provided, against the possibility of a design-basis accident in water-cooled and sodium-cooled reactors, is the coolant.

1. Water Droplets

Under normal operating conditions, reactor off-gases do not contain a significant amount of entrained water droplets.   Under accident conditions in a water-cooled reactor plant, however, it would be necessary to remove very large quantities of water droplets from the air to be cleaned.   Water has a deleterious effect both on the efficiency of HEPA filters for the collection of aerosols and on the efficiency of impregnated charcoal beds for trapping methyl iodide.   Very large quantities can damage or even rupture filters. Water droplets are removed by high-efficiency moisture separators (Section IV-B-1).

2. Comparatively Large Solid Particles

Unlike aerosols, larger particles will not remain airborne long in a quiet, contained atmosphere.   Under normal conditions therefore, relatively few large particles reach air cleaning systems, but prefilters are often provided to remove them in order to minimize the loading of ultrahigh-efficiency HEPA filters.   The prefilters (Section IV-B-2) do not trap aerosols with the high efficiency required for radioactive materials, but they are necessary to prevent clogging of HEPA filters or charcoal beds and possible rupture of HEPA filters.

Comparatively large particles may reach air cleaning systems in considerable quantities in the turbulent containment atmosphere in the first few minutes after a postulated accident, and may contain or adsorb various amounts of radioactive materials. Particles of fuel, other core materials, and corrosion products could all be conveyed through the containment media.

Graded deep-bed filters will be provided in addition to the prefilters in sodium-cooled reactor plants to reduce high concentrations of sodium oxide particles in the event of a major accident.

3. Aerosol Particles

Aerosol particles are particles that are small enough and/or light enough to be kept suspended in air or another gas by Brownian (thermal) motion against the force of gravity. These particles range in size from clusters of a few molecules to particles about 50 microns in diameter. Aerosols produced in a postulated accident may be largely composed of fission products, fuel compounds, and cladding materials that have volatilized from melting fuel and subsequently condensed and agglomerated in the containment atmosphere. They will probably carry water and reactive gases adsorbed on their surfaces. The rate of agglomeration depends largely on particle concentration, particle size, humidity, and electrostatic charges. In reactor applications, aerosol particles are removed predominantly by HEPA filters (Section IV-B-3), although some of them are removed by the prefilters upstream of the HEPA filters. The abilities of containment atmosphere cooling sprays and pressure-suppression pools to collect aerosol particles and larger particles are currently being studied (Section VI-C).

The common industrial problem of removing solid particles from a gas stream is usually solved with conventional equipment such as cyclones, electrostatic precipitators, scrubbers, and filters. The removal of radioactive particles from off-gas streams at nuclear installations is a more difficult problem. Because of the necessarily stringent regulations regarding the dispersion of radioactivity into the atmosphere, high collection efficiency (a way of expressing the removal factor) and reliability are required, and the airborne particles that must be removed include extremely small particles (less than 0.1 micron in diameter). Consequently, filters must have ultrahigh efficiency for submicron particles. Seals, welds, and all structural components must be leaktight. Furthermore, filter systems provided for post-accident use must be rugged in order to maintain their structural integrity, including leaktightness, after a postulated accident. The systems must be designed to protect the delicate ultrahigh-efficiency HEPA filters from shock waves, pressure surges, high moisture loadings, etc. (Section VI).

Most filters will remove a large weight fraction of the particulate matter. However, a high collection efficiency (e.g., 99.99 %), expressed as the weight

TABLE IV

Percentage Fractions by Weight and by Number for an Arbitrary
Particle Size Distribution

Particle diameter (microns)	Relative[a] weight of particle	Number of particles in size fraction	Total relative weight of fraction	Approx. weight fraction (%)	Approx. number fraction (%)
100	10^6	1	10^6	99	0.09
10	10^3	10	10^4	0.99	0.9
1	10^0	100	10^2	0.0099	9
0.1	10^{-3}	1000	10^0	0.000099	90
		1111		99.999999	99.99

[a] Normalized; unit weight assigned to 1 micron particle.

fraction collected regardless of particle size, does not qualify a filter for the removal of submicron particles. Submicron particles accounting for a small fraction of 1 %, by weight, of the particles to be removed from an air stream could account for the great majority by particle count.

For example, assume that the particles in an air stream are all spherical and all have the same density, so that their relative weights are in direct proportion to their relative volumes. Since the volume of a sphere is proportional to the cube of its radius ($V = \frac{4}{3}\pi r^3$) or diameter ($V = \frac{1}{6}\pi d^3$), the weights of the spheres are similarly proportional to the cubes of their diameters. For the arbitrary particle size distribution shown in the first and third columns of Table IV, the percentage fractions by weight and by number would be those given in the last two columns. Here the 0.1-micron particles are less than 10^{-4} % (one part per million) of the total by weight and yet constitute 90 % by particle count. In this case it is mathematically possible for a filter with a collection efficiency of 99.99 %, expressed as a percentage of the total weight of all the particles, to fail to trap any of the 0.1-micron particles (or any of the 1-micron particles for that matter). In practice, although such a filter would trap some of the smallest particles, many would pass through.

If the particle size distribution in Table IV were altered by assigning an equal number of particles to each diameter, the 0.1-micron particles would be less than 10^{-7} % (one part per billion) of the total by weight but would be 25 % by particle count.

4. Iodine and Other Chemically Reactive Gases

The most important chemically reactive gases are molecular iodine (I_2) and two of its gaseous compounds, hydrogen iodide (HI) and methyl iodide (CH_3I). The relative production of these three forms of iodine in an accident cannot yet be quantitatively predicted (39–43), but I_2 will probably be over 90% of the total. When fuel meltdown occurs in an oxidizing atmosphere the iodine is primarily I_2; in a reducing atmosphere it is primarily HI (or solid compounds, such as cesium iodide). Investigations indicate that CH_3I is formed not during meltdown but by subsequent chemical reactions within the containment shell. The relative amounts of the three forms of iodine will depend on the containment atmosphere (air versus inert gas), the amount of hydrogen produced from metal-water reactions, traces of organics in the fuel, and the amounts and types of organics in the containment shell. Methyl iodide and other organic iodine compounds are not likely to constitute more than a few percent of the total (39–42). Nevertheless, all three airborne forms of iodine must be removed as efficiently as possible.

Since iodine and its gaseous compounds are chemically reactive, they can be removed by a variety of physical and chemical processes: adsorption, absorption, chemisorption, solution, etc. The processes that have proven to be most practical and are a part of current reactor plant design are sorption on specially impregnated activated charcoal (Section IV-B-4) and scrubbing with a chemical spray (Section VI-C). Iodine that has been adsorbed by or reacted with particulate material can be removed by HEPA filters.

Other, much less important reactive gases are bromine and trace amounts of other fission-product isotopes with vapor pressures high enough to volatilize them under accident conditions.

5. Noble Gases

Since they are chemically and biologically inactive, noble gases are not presently considered a major hazard except for the contribution of krypton-85 to the cumulative contamination of the atmosphere. Particular attention must be directed to possible exposure to external whole-body radiation on the reactor site if a major accident occurred.

Because of their lack of chemical reactivity and the large volume of air involved, the noble gases xenon and krypton could not be removed, in a practical way, from the containment atmosphere of a power reactor during the high-pressure high-temperature stage of a loss-of-coolant accident (44). The only currently practical method of minimizing the discharge of noble gases to the environment during the early stages of an accident is to retain them in a low-leakage containment system. The maximum acceptable design-basis-accident leakage rates quoted in Section II-B are 0.1 and 0.5% (depending on other safety provisions) of the contained volume per 24 hours (also see Section VI-B-1).

The short-term hazards of external radiation from noble gases after an accident can be minimized by holdup for radioactive decay of the short-lived noble gas isotopes, followed by great dilution and controlled discharge of the long-lived ones through a high stack.   Holdup and stacks are both widely used in normal operations (Section V).

With the problems of submicron particles and methyl iodide largely solved by HEPA filters and specially impregnated charcoal, in recent years an increasing amount of research and development work has been directed to the removal of krypton and xenon from off-gases.   This has not been an urgent problem, but the present practice of releasing long-lived krypton-85 (half-life about 10.3 years) to the outside atmosphere will lead to significant cumulative contamination within the next few decades.   Some of the more promising methods of noble gas removal are listed below.   So far, none of these has been applied in a large power reactor, except holdup on charcoal (the more usual practice is holdup in decay tanks, drums, or large-diameter pipes, as discussed in Section V).

The simplest method for the removal of xenon and krypton is adsorption on charcoal at room temperature.   This requires a very large volume of charcoal. In most cases, the main function of adsorption on charcoal is to delay the release of noble gas fission products long enough for them to decay to acceptable levels.   Adsorption on charcoal or molecular sieves at or below the temperature of liquid nitrogen requires complex equipment of relatively small volume (45,46).

Removal by solution in liquid nitrogen (47) has the advantages of a continuous process but requires large and complex equipment.   Separation by selective permeability in a cascade of membranes (48,49) shows promise but has not yet been proven economically feasible.   Absorption in a liquid fluorocarbon is being evaluated for large-scale application and economic optimization (50).

B. Components of High-Efficiency Air-Cleaning Systems

The air (or other gas) to be cleaned by a high-efficiency system usually passes through system components in this order:

1. High-efficiency moisture separators (if the air may contain enough water droplets to impair the performance of HEPA filters and charcoal).

2. Prefilters (conventional industrial or air conditioning filters)

3. HEPA filters

4. Beds of impregnated activated charcoal

5. Another bank of HEPA filters (to retain fine particles of charcoal swept from the beds by the air stream).

Other components include fans, motors, housings, mounting frames, dampers, isolation valves, and gaskets.

Typical arrangements of high-efficiency systems for various types of reactors and reactor containments are shown in Figure 11 (51). The vented containment for water-cooled reactors mentioned in the figure is used for many research reactors and for USAEC production reactors (51). Vented containment is not used for water-cooled power reactors, because of their large size and the economic desirability of siting them close to power consumers to reduce power transmission costs.

Air-cleaning systems for pressurized-water and boiling-water reactors are shown in Figures 6 and 7, and typical component arrangements for these reactors are shown in Figures 12 and 13. In pressurized-water reactors, the charcoal beds in the recirculating system are usually bypassed during normal operations (Figures 6, 11, and 12). In boiling-water reactors, the moisture separators and charcoal beds are usually part of a separate high-efficiency air-cleaning system held in standby during normal operations (Figures 7, 11, and 13).

An air-cleaning system for a Magnox gas-cooled reactor is shown schematically in Figure 8 (also see Figure 11). Components of the air cleaning system for a helium-cooled reactor are not specified in the schematic drawing of its containment (Figure 9) but will be similar to those for a Magnox reactor. The gas cleanup system for the helium coolant circuit contains charcoal beds and liquid nitrogen traps (33).

The arrangement for a sodium-cooled reactor is shown in Figures 10 and 11. This system will probably not need charcoal or moisture separators. In systems provided for post-accident air cleaning in sodium-cooled fast breeder reactors, the prefilters will quite possibly be preceded by baffles and graded deep-bed filters for the removal of sodium oxide aerosols. Many air-cleaning systems for power reactors include large drums (Figure 6) or sections of large-diameter piping (Figure 7) to provide on-stream volumes to delay the passage of the air or gas long enough for short-lived radioactive gases to decay, in large part, to nonradioactive materials. Quite a few reactors, particularly boiling-water reactors, have tall stacks to disperse routinely released radioactivity high above ground level, although releases during normal operations are within acceptable limits. In fact, most reactor plants have released only a small fraction of their acceptable limit.

Pressurized-water reactors, which have low rates of flow from the reactor primary system during normal operations, have tanks in which off-gases can be stored if necessary until meteorological conditions (wind direction and velocity, lack of atmospheric inversion, etc.) are favorable for the proper dispersion of long-lived Krypton-85.

The four major components of high-efficiency air-cleaning systems are described and discussed briefly in the following subsections, with emphasis on HEPA filters and beds of specially impregnated charcoal. Detailed

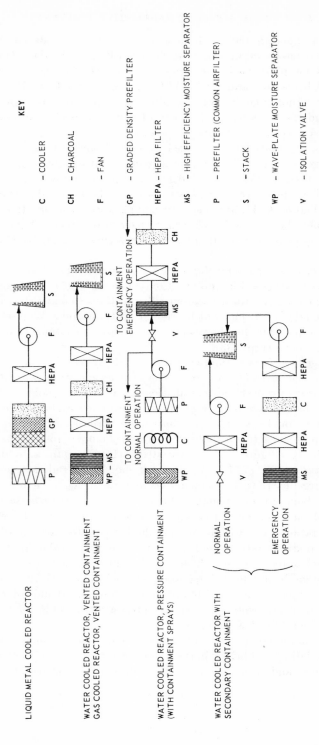

Fig. 11. Arrangements of air-cleaning systems for various types of reactors and reactor containments (51).

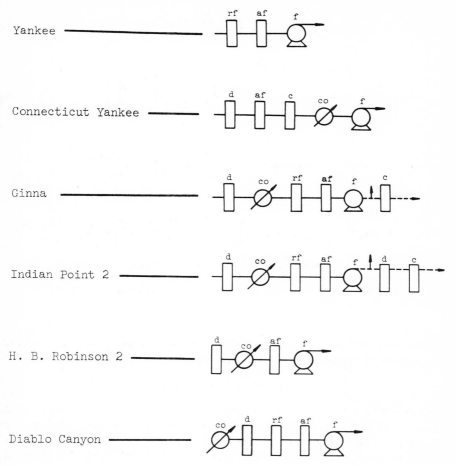

Fig. 12. Typical component arrangements for recirculating air-cleaning systems within containment shells (d = high efficiency moisture deentrainer; rf = roughing filter or prefilter; af = HEPA filter; c = charcoal adsorber; co = cooler; f = fan; solid lines indicate normal flow; dashed lines indicate accident flow).

descriptions of the design, construction, and testing of high-efficiency air-cleaning systems and their components are contained in (51), which is an excellent, practical, up-to-date handbook addressed primarily to designers and architect-engineers, but highly recommended to any reader of this chapter who needs authoritative, detailed, practical information. This handbook reflects the experience of users of high-efficiency air-cleaning systems in power reactor plants, USAEC reactor installations, reactor fuel

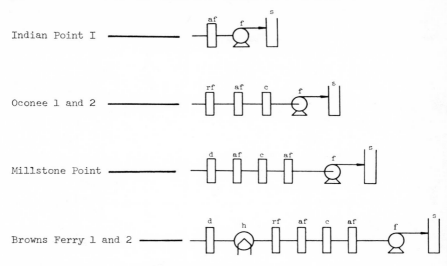

Fig. 13. Typical component arrangements for once-through air-cleaning systems in secondary containment exhaust lines (d = high efficiency moisture deentrainer; af = HEPA filter; c = charcoal adsorber; rf = roughing filter or prefilter; f = fan; s = stack; h = heater).

manufacturing plants, and other nuclear installations, as well as the experience of component designers and manufacturers.

1. Moisture Separators

Air-cleaning systems provided for removal of fission products after a postulated accident involving loss of coolant, in a water-cooled power reactor, or involving rupture of a heat exchanger, in a gas-cooled power reactor, require moisture separators upstream of HEPA filters and charcoal beds. The atmosphere entering air-cleaning systems each minute after such accidents may contain several pounds of water droplets per filter. High moisture loadings can reduce the ability of the HEPA filter media to withstand pressure differentials and pressure surges. Even higher loadings can plug the filters, causing rapid rupture. Furthermore, flooding of charcoal beds can reduce their efficiency for trapping elemental iodine and drastically reduce their efficiency for trapping methyl iodide. The effectiveness of moisture separators in protecting HEPA filters and charcoal beds is illustrated by the following descriptions of tests on two types, one of which incorporates a wave-plate separator for removal of the larger drops.

Extensive tests have been made of the ability of a commercial moisture separator to protect HEPA filters and activated charcoal beds after a postulated loss-of-coolant accident (52,53). The separator was a York M321 Demister,* which contains about 24 woven mats in series. The mats themselves are Teflon† yarn woven on stainless steel wire and wrapped with stainless steel reinforcing wire. The complete Demister has a 24 × 24 in. face, is 2 in. thick, and is rated for 1600 scfm at 0.95 in. W.G. The filters used in the test were 1000-cfm open-face steel-cased HEPA filters, 24 × 24 × 11½ in. The tests showed that when moisture separators were provided, the HEPA filters successfully passed flows of wet steam and fog mixtures for 10 days, even when the filters contained loadings of normal dust which were equivalent to 18 months of exposure to flowing air (52). These tests simulated a power-surge accident, followed by evaporation of the coolant. In a simulation of failure to reestablish coolant flow, wet steam was emitted for 30 sec at the rate of 7000 scfm per filter, and the subsequent evaporation caused mixtures of steam, air, and entrained water to be evolved at rates up to 1 lb/min per filter for 10 days (53). In a typical test, partial plugging of the dust filters by water particles that initially escaped the separator reduced the mixed flow to a minimum of 600 cfm 6 hr after the test started; however, the flow gradually increased to over 900 cfm at the end of the test. After the test, the HEPA filters exceeded their efficiency requirement. On the other hand, filters tested without moisture separators upstream were almost completely plugged with liquid water, and some ruptured during the fog test.

Tests on another moisture separator were conducted (54) to determine its effect on entrained water from containment atmosphere cooling sprays. This separator consists of wave-plate steel baffles followed by three 2-in.-thick fiber glass pads, with a space of about 1 in. between the second and third pad. The test facility used was capable of circulating 1000 cfm of saturated air-steam mixtures at as high as 40 psig and 261° F, the maximum predicted conditions. Water sprays simulated the droplet cloud expected. The moisture separator removed essentially all droplets from the air stream under six test conditions with pressures ranging from 10 to 40 psi. HEPA filters downstream of the separator (1000-cfm filters, tested one at a time) withstood test periods of up to 24 hr at predicted maximum conditions without measurable loss of performance.

2. Prefilters (Roughing Filters)

Prefilters (roughing filters) are ordinary filters of the types commonly used in air conditioning and industry. They are often needed to protect

---

* Trade name of Otto H. York Company, Inc.

† Trade name of E. I. du Pont de Nemours and Company, Inc. for polytetrafluoro-ethylene.

TABLE V

Classification, Removal Efficiencies, and Prices of Common Air Filters
and HEPA Filters (Modified from (51))

Group I filters are viscous impingement, panel type; Groups II and III filters and HEPA
filters are extended-medium, dry type

Group	Efficiency	NBS efficiency[a] (%)	Removal efficiency (%) for particle size of				Price range per 1000 cfm (dollars)
			0.3	1.0 (micron)	5.0	10.0	
I	Low	5–35[b]	0–2	10–30	40–70	90–98	1–20
II	Moderate	40–75[b]	10–40	40–70	85–95	98–99	25–40
III	High	80–98[c]	45–84	75–99	99–99.9	99.9	30–65
HEPA	Extreme	100[c]	99.97 min	99.99	100	100	60–80[d]

[a] National Bureau of Standards, Dill Dust-Spot Method (R. S. Dill, A Test Method for Air Filters, National Bureau of Standards (1938)).

[b] Test using synthetic dust.

[c] Test using atmospheric dust.

[d] Unit cost of 1000 cfm filter.

HEPA filters from comparatively large particles and from overloading. In some installations, HEPA filters in systems provided to clean reactor off-gases or air in exhaust systems are preceded by prefilters within the high-efficiency systems. In other installations, the HEPA filters in the exhaust air-cleaning system are protected by prefilters in the building air intake system. Prefilters are necessary in all high-efficiency systems provided for cleanup of the containment atmosphere after a postulated accident.

In general, HEPA filters should be protected from particles larger than 1 or 2 microns in diameter, from lint, and from dust concentrations higher than 10 grains per 1000 ft^3. However, when the radioactivity of dust collected on HEPA filters is high enough that they must be changed weekly or monthly, the extra cost of prefilters may not be justifiable (51).

The classification, efficiencies, and prices of prefilters used in the United States are shown in Table V (modified from (51)); HEPA filters are included for comparison. Groups II and III filters (and HEPA filters) are extended-medium filters; the medium is pleated or formed as bags or "socks" to give large surface area with minimum frontal area (51). Representative air-flow capacities, resistances, and dust-holding capacities are given in Table VI (also modified from (51)). Dust-holding capacity under various service conditions cannot be accurately predicted on the basis of laboratory tests or manufacturer's catalog data; Table VI is presented for comparison only (51).

3. High-Efficiency Particulate Air (HEPA) Filters

The term HEPA filter is much more specific than would be inferred from its derivation (the acronym for "high-efficiency particulate air filter"). HEPA filters constitute a highly useful and very specialized type, the culmination of about 40 years of development, theoretical and experimental research, and experience by both government and industry in a number of countries (55).

HEPA filters have an ultrahigh collection efficiency. Furthermore, this is on the basis of the number fraction of submicron particles collected, not on the basis of weight fraction of particles collected. HEPA filters used in the Atomic Energy Program in the United States, for example, are required to test at not less than 99.97 % efficiency for the retention of particles of a standard test aerosol with a uniform particle size, 0.3 micron in diameter (Section VII-A-1). This was considered the most penetrating particle size, based on pioneer high-efficiency filtration theory in 1942; the most penetrating particle size for HEPA filters with glass media is presently considered to be about 0.07 micron, as discussed in Section VII-A. Efficiency is the conventional term for the percentage removed, by the filter, of the quantity of a test aerosol in an air stream introduced into the filter. Penetration is the term for the percentage that passes through the filter.

In most HEPA filters used in reactor plants today the medium is a paper made of super-fine spun glass of two or more fiber diameters. Filtering fibers, as fine as 0.25 micron in diameter, are mixed with larger support fibers. Glass-asbestos paper is also used. HEPA filters are given a very large area of medium per unit face area of the filter by deeply pleating the medium and inserting separators on each side of each pleat to prevent sagging or collapse of the medium. The separators are corrugated to allow the passage of air. There are about three complete pleats and six separators

TABLE VI

Air Flow Capacity, Resistance, and Dust-Holding Capacity of
Air Filters (Modified from (51))

Group	Efficiency	Air flow capacity (cfm per sq. ft. of frontal area)	Resistance (in. H_2O) Clean filter	Used filter	Dust-holding capacity (lb per 1000 cfm of air flow capacity)
I	Low	300–500	0.05–0.1	0.4–0.3	1–3
II	Moderate	250–750	0.1 –0.5	0.2–0.5	1–5
III	High	250–750	0.20–0.5	0.6–1.4	1–5
HEPA	Extreme	250	1.0 max.		4

to the inch. The most widely used HEPA filter has a 24 × 24 in. face and a depth of $11\frac{1}{2}$ in. Although its superficial face area is only 4 ft², it contains at least 200 ft² of medium (usually closer to 250 ft²) because of the deep pleating. This enables a filter with a flow of only 5 fpm through the glass fiber paper to accommodate a flow of at least 1000 cfm in the 4 ft² of face area.

The efficiency, air-flow capacity, resistance, dust-holding capacity, and approximate price range (for 1970) of HEPA filters are listed in Tables V and VI.

The top and bottom edges of the medium and the separators and the ends of the medium are all sealed to the inside of the frame with an adhesive. The construction of HEPA filters is shown in Figure 14 (51). Steel frames are recommended for use in systems provided for post-accident cleanup; wood frames, which are more resistant to vibration, are recommended for other purposes (56).

Each HEPA filter must be tested and inspected before installation and must be packaged, shipped, handled, and installed with care. HEPA filters must be routinely tested in place as part of the filter system, and individually when necessary, throughout their service life. Also, system leakage that allows air to bypass the filters must be extremely low. HEPA filter systems

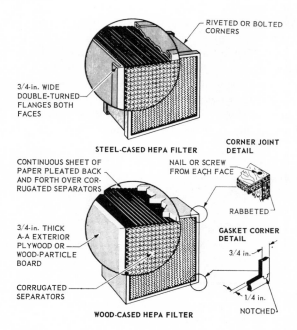

Fig. 14. Construction of open-face HEPA filter units (51).

must be designed, tested, inspected, and maintained so that the conditions under which the filters themselves operate are good. The filters must be protected from shock waves, pressure surges, high pressure differentials, excessive moisture and particulate loadings, and fire.

4. Specially Impregnated Charcoal Beds

Beds of activated charcoal have high efficiency for the trapping of radio-active molecular iodine ($^{131}I_2$), provided the charcoal has not lost efficiency from flooding or from long exposure to moisture or impurities in the air; has not settled to the extent that air bypasses the charcoal through leakage paths; and does not become ignited by the decay heat of accumulated fission products; and provided that it is not extensively wetted or flooded with water during any time that air or other gas containing iodine is actually entering the beds.

Activated charcoals, however, do not remove methyl iodide or other organic iodides efficiently at high humidities, unless they are specially impregnated for that purpose. Unimpregnated activated charcoals would not be adequate for the post-accident removal of methyl iodide in recirculating air-cleaning systems within the containment shell of water-cooled power reactors, nor would they be adequate in a gas-cooled power reactor accident if a heat exchanger failed. Unimpregnated charcoals are still in use in some reactor air-cleaning systems; some of these charcoals may have adequate efficiencies for the removal of methyl iodide in filter-adsorber systems in the exhaust of secondary containment structures, but only if it can be assured that humidity can be reduced to very low levels. Further, the amount required is much larger than the amount of impregnated charcoal required for the same application. Therefore, unimpregnated charcoal is no longer recommended for use in power reactor containment systems.

The formation, reactions, and transport of methyl iodide and other organic iodides within the containment system of a power reactor during and after an accident would depend on many parameters. The most important of these are the nature of the containment atmosphere, reactions of the methyl iodide with bare or painted surfaces, temperatures in the system, iodine concentrations, and the ratio of the total surface area within the containment system to its volume. The amount of iodine that would reach the charcoal beds in the form of methyl iodide or other organic iodides cannot yet be quantitatively predicted, but it is considered to be a small percentage (0.5 to 5%) of the total iodine inventory and a somewhat larger percentage (possibly 10%) of the airborne iodine (40,42,57–59). However, it is generally assumed that this amount of methyl iodide would be significant (60–62). Therefore, for siting reactors in or near population centers, an assurance of a high-efficiency for the removal of methyl iodide may be required.

In the United States, the means provided for preventing the release of radioactive methyl iodide (CH_3 ^{131}I) after a postulated loss-of-coolant accident in a water-cooled power reactor is removal of the radioactive iodine from the methyl iodide by isotopic exchange. For this purpose, activated coconut charcoal is impregnated with one or more substances containing nonradioactive iodine ([127]I), which exchanges with the radioactive ^{131}I in the methyl iodide as the air being cleaned passes through the charcoal bed. In this manner the impregnated charcoal removes and traps the radioactive ^{131}I and releases nonradioactive CH_3 [127]I. Some methyl iodide, as such, may be trapped by the impregnated charcoal, but the main trapping mechanism is the isotopic exchange.

Research and development work on charcoals specifically impregnated for removal of iodine-131 from methyl iodide is intensive, and has been for several years. Several types of impregnated charcoal have been found to be effective. Undoubtedly, more will be developed in the near future. The reader should not receive the impression that the type discussed in the next four paragraphs is preferred over other types produced by the same or other manufacturers, provided that they have been shown to be effective. Specific application, availability, and other factors must be considered. The advice of an expert should be obtained during the design stage. With these cautions in mind, the impregnated charcoal referred to below, Mine Safety Appliances Type MSA-85851 (8 × 14 Tyler mesh) is highly recommended.

A commercially manufactured and impregnated charcoal, type MSA-85851, has been shown to be effective for removing radioactive iodine-131 from CH_3 ^{131}I at the approximate temperature, pressure, and humidity predicted for the atmosphere within the containment shell of a water-cooled power reactor after a postulated loss-of-coolant accident. This is subject to the provision that wetting or flooding of the charcoal by condensed steam can be prevented. The effectiveness of this charcoal was investigated in a series of 18 tests (63) conducted for Westinghouse Electric Corporation, Nuclear Energy Systems Division, by Oak Ridge National Laboratory, with the approval of the USAEC. The costs were borne by Westinghouse. The approximate test conditions were 130° C (266° F), 55 psia, 40 ft/min steam-air face velocity, and a calculated average relative humidity of 94% or higher. The inlet concentration of methyl iodide in the flowing steam-air mixture was 6 mg/m^3 (except for two tests with 80 mg/m^3). The charcoal beds were 3.5 in. in diameter and 2 in. deep. Residence time (duration of contact of CH_3I with charcoal) was 0.25 sec. Two modes of flow, downward and upward, were used. In all but two tests, the charcoal was retained in a cylinder between disks of perforated metal of the type used in charcoal beds for reactor installations.

Six of the 18 tests were performed with charcoal that had been flooded with water and then purged with a steam-air mixture. Of the 12 tests performed without preflooding, two were performed with CH_3I inlet concentrations of 80 mg/m³ and two others were performed with stainless steel wire screen instead of the perforated metal. The 8 basic tests were conducted with a CH_3I inlet concentration of 6 mg/m³, with perforated metal retainers, and without preflooding; the amount of CH_3I injected was 0.13 mg/cm³ of charcoal (this is approximately 0.30 mg/g of charcoal, or 300 μg/g).

The results of the eight basic tests are shown in Figure 15 (63). The note

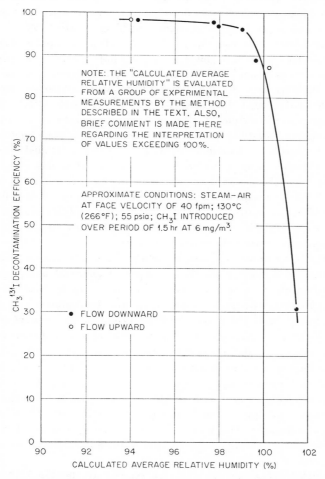

Fig. 15. Effect of relative humidity on the efficiency of MSA 85851 iodized charcoal, lot no. 021369, for decontaminating radioactive methyl iodide transported in flowing steam-air (63).

on the figure, regarding the method of evaluating the calculated average relative humidity, refers to the text of (63). The calculated average relative humidity for each test is based on the average temperature, pressure, and composition of the steam-air mixture over, approximately, the period during which CH_3I was injected. The range of probable error of the calculated average relative humidity was estimated at ± 1.0 to 1.5%. The actual relative humidity in the charcoal test bed varied axially and possibly radially, and in most cases the tests were conducted in the vicinity of saturation conditions. In tests where the calculated average relative humidity exceeded 100%, a small part of the flowing steam was transported through at least a portion of the test bed in a condensed phase (water) (63).

From these tests it was concluded (63,64) that this charcoal is effective at relative humidities as high as 99% provided that certain other conditions are not too severe, as discussed below, and provided that certain circumstances such as preflooding are avoided. Decontamination efficiency (ability to remove iodine-131 from CH_3I) is relatively insensitive to CH_3I inlet concentration over a wide concentration range. If operating conditions are such that relative humidity tends to exceed 100% in some portion of a charcoal bed, charcoal in that portion will be wetted or flooded and have a low $CH_3\ ^{131}I$ decontamination efficiency. Charcoal that has been flooded and then purged with saturated air before testing has a low decontamination efficiency with downward flow and a lower efficiency with upward flow. In the absence of extensive wetting or flooding, direction of flow does not affect efficiency. Flooding with condensed steam results in significant loss of available impregnant from the charcoal granules.

The charcoals recommended for use in the United States contain about 5 wt. $\%$ impregnating material. These types of charcoal are proprietary, and little information on their characteristics is available. However, they are known to contain nonradioactive iodine and to function primarily by isotopic exchange, and they have useful efficiencies for methyl iodide at loadings of about 300 μg of methyl iodide per gram of charcoal compared with only about 100 μg (under different conditions; see the next paragraph) for charcoal impregnated with 0.5 wt. $\%$ KI. Therefore, the addition of a complementary charcoal, such as one impregnated with triethylenediamine, is not recommended for these charcoals for use in the United States. One of the requirements for impregnated charcoal in the United States is that it must perform efficiently with relatively large quantities of CH_3I under the postulated steam-air conditions of a loss-of-coolant accident in a water-cooled reactor.

In the United Kingdom, in intensive studies at the UKAEA Reactor Development Laboratory at Windscale, which included tests of a number of inorganic and organic impregnants, a mixture of two impregnated charcoals gave the best results (65). Both of these were prepared from the same coal-base charcoal. One was impregnated with 0.5 wt. % potassium iodide (KI); the other was impregnated with 5 wt. % triethylenediamine (TEDA). (Unlike the iodized charcoals, TEDA-impregnated charcoal traps CH_3I as such). For KI-impregnated charcoal, the best performance in the Windscale tests was considered to be in the range 0.1 to 1 wt. % KI; 0.5 wt. % was selected as being in the middle of that range. However, the charcoal impregnated with 0.5 wt. % KI was found to have an acceptable methyl iodide loading of only slightly over 100 $\mu g/g$. Serious overloading of the charcoal impregnated with 5 wt. % TEDA did not occur until it was loaded with well over 1 mg/g, but tests indicated that TEDA was much more subject to loss by volatilization at high temperatures than was KI. Also, in the carbon dioxide used as coolant in most British power reactors, TEDA required more than twice the residence time to give the same performance as in air. These (and other, less important) findings that each of the two impregnants had advantages over the other led to the recommendation (65) of a mixture of the two impregnated charcoals for use in the United Kingdom. It should be noted that the methyl iodide loading requirement is less than that for the conditions in the United States.

In recent experiments, (66) both in the laboratory and in the air within a research reactor building, three types of iodized charcoal were observed to lose efficiency slowly on exposure to flowing air at about 50 % relative humidity. For a 2 in. bed depth, loss in efficiency for the removal of iodine-131 from CH_3I ranged from 0.11 to 1.1 %/month. Frequent in-place testing of beds of iodized charcoal that are to be left on stream was recommended. (As mentioned in the introductory portion of Section IV-B, charcoal beds provided for post-accident use are in standby in most water-cooled reactor plants during normal operations). Charcoal impregnated with 5 % TEDA was also tested, and lost very little of its efficiency for trapping $CH_3\,^{131}I$ (66).

Beds of activated charcoal must be designed to prevent or counteract settling. Leaks that bypass air around the charcoal can drastically reduce or destroy the capability of charcoal beds. This is also true of leakage caused by channeling in charcoal beds extensively flooded with water. Activated charcoals, particularly impregnated charcoals, must be protected against ignition by fission-product decay heat. Furthermore, impregnated charcoal will release some of its impregnant and may release trapped fission-product iodine and iodine compounds if subjected to excessive heat well below its ignition temperature. The gradual decrease in efficiency from long-term exposure to flowing humid air is accelerated if the air contains significant amounts of impurities, particularly paint or solvents. For the reasons outlined above, charcoal beds that are depended upon to perform reliably after a postulated accident must, like HEPA filters, be protected against damage or loss of function by well designed and carefully maintained systems.

Two types of charcoal bed adsorbers are generally used in reactor installations in the United States: unit-tray and pleated-bed adsorbers. Unit-tray adsorbers consist of two charcoal beds with an air space between. The adsorber shown in Figure 16 (51) has two $22\frac{1}{2} \times 26 \times 2$ in. beds of 8×16 mesh charcoal and has a capacity of 333 cfm. The contact time between the gases and the charcoal (residence time) is approximately 0.24 sec at rated flow. Depending on minor variations in designs, unit-tray adsorbers cost (1970) from about $350.00 to $400.00 (with stainless steel frames) and weigh from 80 to 100 lb. Some models have 12×30 mesh carbon and therefore require cotton scrims inside the screens. Cotton scrims are not recommended for nuclear exhaust applications, because they could burn out or deteriorate, permitting the release of fine charcoal particles. Some models have caulked joints between the screens and the bed casings and between the individual beds and the outer casing. The design does not lend itself to recharging with new carbon. This design is preferred by most power reactor system designers

Fig. 16. Unit-tray charcoal adsorber. Note through bolts within the screen area (51).

because of its compactness and the ability to get the full 2 in. depth of carbon in a single unit (51).

A more conservative adsorber design is the pleated-bed type, shown in Figure 17 (51). This design is favored by contractors who operate AEC-owned reactors because there are no internal through bolts, spacers, or caulked joints; because the units fit the standard $24 \times 24 \times 11\frac{1}{2}$ in. module; and because the units are light enough (160 lb) to be handled by two men. The unit shown has a 1 in. bed of 8×16 mesh carbon and a capacity of 800 cfm at 0.8 in. H_2O pressure drop. The contact time between the gases and the carbon at 800 cfm is slightly over 0.1 sec (two units in series are required for methyl iodide). Two-inch-bed models are also available. The cost of the unit shown is about \$400.00 with a stainless steel case or \$325.00 with a carbon steel (painted) case. This type of bed has no internal metal-to-metal interfaces (note external bed spacers), no caulked joints, and no scrims.

Fig. 17. Pleated-bed charcoal adsorber. End plate removed to show charcoal bed (51).

The end plates, which are sealed with highly compressed neoprene pads, can be removed for recharging with new charcoal (51).

V. TREATMENT OF REACTOR OFF-GASES DURING NORMAL OPERATIONS

There are a number of similarities between the treatment of airborne radioactive wastes during normal operations and the provisions for cleanup of containment air after postulated accidents. In particular, stringent regulations are imposed on the amounts and concentrations that can be released to the outside atmosphere, and high-efficiency air-cleaning systems and tall stacks are used wherever necessary. Air-cleaning systems and their components must be designed, fabricated, tested, and maintained for continuing effectiveness and reliability.

On the other hand, normal treatment of airborne radioactive wastes differs in a number of very significant respects from provisions for post-accident cleanup. During normal operations the following conditions apply. The amounts of radioactive fission products that become airborne or gasborne are so low that neutron activation products constitute a large part of the airborne and gasborne radioactive isotopes. Off-gases for treatment are collected directly from the primary coolant system and are not diluted with containment ventilation air until they have passed through the air-cleaning system. The only fission products that enter the free space within the containment are the small amounts that leak through seals, valve stems, etc. in components of the primary coolant system and the primary coolant treatment systems. This is generally also true of airborne activation products. Air-cleaning systems are required to treat only normal flows of air or other gas at normal pressures, temperatures, and humidities, and particulate loadings are very low.

Releases of iodine-131 and particulate radioactive materials have been so low during normal operations that the less critical problem of minimizing the cumulative contribution to the earth's atmospheric radioactivity over future decades by long-lived krypton-85 (10.3 year half-life) is receiving an ever increasing amount of attention. As a matter of fact, the quantity of radioactivity of halogens (principally iodine-131) and particulates released to the environment from water-cooled reactor plants in the United States has been only about 10^{-6} as high as that of noble and activation gases released (38,67).

As seen in Section III, by far the most important airborne radioactive materials produced by reactors during normal operations are the fission products krypton-85 and iodine-131 and the neutron activation product argon-41. Iodine can be trapped with high efficiency by beds of impregnated charcoal during normal operations, but at present krypton and argon, being chemically inactive noble gases, cannot be separated from other gaseous effluents by economically feasible methods. It is possible to bottle the off-gas from the primary coolant system of a pressurized-water reactor, but the volume of off-gas flow from a boiling-water reactor is too high for this. The volumetric flow rate of the shield cooling air of a Magnox gas-cooled reactor is too high for economical use of holdup drums or tanks for the decay of argon-41.

Fortunately, in almost all cases, releases through the years have not exceeded a few percent of the conservatively low acceptable levels (29,67–69). Therefore, it has been possible to practice continual controlled release of airborne radioactive noble gases in compatibility with nuclear safety requirements. Air-cleaning requirements are continuously followed by monitoring the effluent gases in the stack or vent and the atmosphere in the outside environment. (Monitors are also used at various points within the reactor system.)

A. Treatment in Water-Cooled Reactor Plants

In pressurized-water reactors, the primary coolant circulates in a closed system between the reactor core and the heat exchanger (see Figure 6). Since the primary coolant does not drive the turbine or pass through a condenser, essentially no air is introduced into the primary coolant system during normal operations. Normally the primary coolant is supersaturated with hydrogen to suppress radiolytic decomposition of the water. In some plants the hydrogen is reacted with oxygen in a catalytic combiner, forming water which enters the liquid waste system. In others the hydrogen is purged with nitrogen. In any case, off-gas volumes are low enough that the gases can be stored for weeks for radioactive decay (after being compressed if necessary). All pressurized-water reactors have storage drums or tanks for this purpose. In some of these plants, however, such storage has not been used and has proved unnecessary. Off-gases from bypass coolant purification systems, pressure-relief tank vents, drain tank vents, and sampling stations are usually transferred into surge tanks (67) and may subsequently be stored in drums or tanks for decay of short-lived radioisotopes. Eventually the gases are passed through HEPA filters and, if necessary, through impregnated charcoal beds. Finally, the treated gases are diluted with containment ventilation air (50,000 to 300,000 ft^3/min (38); 70,000 for the Connecticut Yankee (67), for example) and discharged through vents or stacks. The containment ventilation air is continuously recirculated through moisture separators, prefilters, and HEPA filters, with charcoal beds on standby for removal of iodine and methyl iodide if necessary.

Tall stacks for dispersion of effluent air far above ground level have been unnecessary for pressurized-water reactors, as far as normal releases are concerned. However, tall stacks may be considered useful in mitigating the consequences of postulated accidents in all large power reactors.

In boiling-water reactors, leakage of air into the turbine condenser (see Figure 7) is usually of the order of tens of cubic feet per minute (67). Therefore, only a short holdup time (about 30 min.) for decay of short-lived isotopes is practical, and long-term storage of off-gases is not feasible. The volume can be reduced by catalytic recombination of hydrogen and oxygen. Off-gases from the main condenser air ejector are delayed for a few minutes, then exhausted through HEPA filters and a tall stack. Off-gases from the turbine gland seals are also exhausted through HEPA filters and the stack, but after a delay of only 1 or 2 min. Impregnated charcoal is used to remove iodine, when necessary.

B. Treatment in Gas-Cooled Reactor Plants

In Magnox gas-cooled reactors, the carbon dioxide coolant is continuously circulated through cyclones, ceramic "candle" filters, or sintered-metal filters to remove particulate activity. The gas circuit is subjected to controlled purging to relieve slight over-pressures and to control chemical contaminants, and the seal oil is degassed routinely. Occasionally a boiler or the whole system is blown down for operational reasons. Blowdown is through high-efficiency filters and impregnated charcoal beds (see Figure 8). The shield cooling air is maintained at a slight negative pressure and is exhausted through bonded, oil-impregnated glass fiber filters. Generally, it has not been necessary to use extra-high-efficiency filters or tall stacks (29).

Helium-cooled reactors have gas-cleaning systems, containing liquid nitrogen traps as well as charcoal beds, in the helium circuit (26). The air or nitrogen in the secondary (or tertiary) containment is exhausted as necessary through HEPA filters and charcoal beds (Figure 9).

C. Treatment in Liquid-Metal-Cooled Fast Breeder Reactor Plants

In liquid-metal-cooled fast breeder reactors, the sodium coolant will be cleaned by cold traps or other devices in bypass lines. Cold traps reduce the temperature of the circulating molten coolant enough to solidify some of the sodium oxides, sodium carbides, and other impurities with higher melting points than that of sodium. With the exception of noble gases, practically all of the radioactive fission-product and activation-product material will probably be retained in the sodium, plated out in the coolant system, or removed in the cold traps. Essentially all of the iodine (and bromine) will be in the form of metallic salts, primarily sodium iodide (and bromide).

Most of the krypton-85 and xenon-133 will diffuse into the argon (or helium) cover gas above the liquid sodium (see Figure 10), accompanied by short-lived noble gas isotopes and small amounts of other radioactive gases. Oxygen will be removed from the cover gas by chemical reaction, and sodium vapor will be removed by condensation in heat exchangers. The cover gas will be replaced slowly and continuously by inleakage of fresh argon (or helium) from seals and other pressurized components. Before refueling, the cover gas will be purged thoroughly. Waste cover gas will be compressed if necessary, stored for several days for decay of xenon-133, and then released gradually through the stack.

The nitrogen in the pressure containment surrounding the reactor vessel and primary coolant system will be filtered and will be purged and released gradually when necessary. HEPA filters preceded by ordinary prefilters should be adequate for normal operations. Charcoal beds may be unnecessary because of retention of iodine as sodium iodide (36).

VI. CLEANUP OF CONTAINMENT ATMOSPHERES AFTER POSTULATED ACCIDENTS

Reactor containment systems are designed so that, if necessary, the containment shell could be kept sealed indefinitely after a major accident, provided the design pressure and leakage rate were not exceeded. As mentioned in Section II, the maximum acceptable design-basis-accident leakage rates for water-cooled reactors are 0.1 wt. % of the contained volume per 24 hr for pressurized-water reactors and 0.5 wt. % of the contained volume per 24 hr (leakage is into a secondary containment building) for boiling-water reactors. These leakage rates are specified at maximum design-basis-accident temperatures (about 275 to 290° F) and pressures (about 40 to 60 psig). Leakage rates specified for gas-cooled and sodium-cooled reactors are of the same order as those for water-cooled reactors; pressures are somewhat lower, because lower pressures are expected in the outer containment than in that of a water-cooled reactor in the event of a major accident.

There is some possibility that the design pressure or the design leakage rate could be exceeded as a result of a postulated accident. These conditions can be minimized or eliminated by high-efficiency air-cleaning systems that clean the containment atmosphere and exhaust it to the outside. The release of airborne radioactive materials during the high-pressure post-accident period can also be reduced by recirculating high-efficiency air cleaning systems or chemical sprays that reduce the amounts of airborne radioactive materials available for leakage.

The probability of a major accident in a nuclear power reactor, as outlined in the introduction, is very low; extensive precautions are taken in design, licensing, construction, inspection, operation, and administrative control to prevent such an accident; and vigorous emergency measures would be taken in the event of a major accident to limit its consequences.   Therefore, the maximum allowable post-accident releases of radioactive materials used as bases in reactor siting calculations are far higher than the releases permissible under normal operating conditions (3–6)   However, continuing efforts are made, not only to further reduce the possibility of a major accident, but also to further reduce the amount of radioactive material that could be released as a result of an accident.   Furthermore, the closer a reactor is to a population center, the larger the center, and the higher the population density, the more restrictive must be the limits on the amount of fission products, fuel material, and activation products that could be released from the containment system.   This applies to design-basis accidents and other major accidents as well as to lesser accidents and normal operations.   In order to implement these restrictions, air-cleaning systems provided for cleanup of the air within the containment after a postulated accident must be designed, constructed, inspected, tested, and maintained so that they will perform effectively and reliably under any conditions to which they might be subjected.

A. Types of Post-Accident Containment Air-Cleanup Systems

Air cleaning is depended on in two major ways to minimize the release of radioactive materials to the environment after a postulated accident: cleaning of containment air as it is exhausted to the outside, and rapid reduction of the concentration of airborne radioactive materials within the containment atmosphere.

For post-accident cleanup of the air to be exhausted to the environment, once-through high-efficiency systems are almost universally provided. Airborne radioactive materials are collected and retained in these systems as the air is exhausted through them in a single pass.   These systems are usually on standby during normal operations, and are provided to clean and exhaust the air from a secondary containment building or volume in the event of an accident, maintaining a slight partial vacuum so that any leakage through the outer walls of the secondary containment would be inleakage from the outside atmosphere.   Once-through systems would also, when necessary, relieve pressure buildup by venting the air or other gas from the inner containment.

For internal post-accident cleanup of air within the containment volume, recirculating systems are provided.   These systems are designed to reduce the amount of radioactive materials available for leakage by rapidly reducing the concentration of airborne fission products, activation products, and fuel material within the containment atmosphere after an accident.   Recirculating high-efficiency air-cleaning systems have been provided for this purpose in a number of pressurized-water reactors.   However, because of the difficulty of ensuring their post-accident integrity and performance, their adequacy as post-accident cleanup systems is questionable.   A recently developed chemical spray system is being provided for internal post-accident air cleanup in pressurized-water reactors.   The spray system would be actuated only after a major accident, and would repeatedly spray the whole volume inside the containment system, recirculating the cleaning spray solution rather than the air to be cleaned.

The post-accident cleanup performance of a once-through system depends almost entirely on its efficiency and integrity.   These systems generally exhaust cleaned air (after holdup time for decay of noble gases) through a high stack into the outside atmosphere.   Therefore they must, and do, have high efficiencies for the removal of particulate and gaseous radioactive materials during a single pass through the system.

The post-accident cleanup performance of a recirculating system depends not only on the efficiency and integrity of the system, but also on the duration of the release of airborne radioactive materials into the containment shell, the degree of mixing which occurs within the containment shell, the number of air changes per hour through the recirculating system, and the length of time available for the system to clean the air.   (For this purpose, air changes per hour is defined as the ratio of the volume of containment air that passes through the air-cleaning system per unit time to the total volume of air within the containment shell.   Typical values are given in Section VI-B-2.) Efficiency per pass in a recirculating system is far less important than the total efficiency of the system integrated over the critical period, just after a major accident, during which cleanup must be effected.   This period has been estimated to be as short as $\frac{1}{2}$ hr under some possible conditions (70).

Air-cleaning systems that clean the off-gas (water-cooled reactors), coolant gas (gas-cooled reactors), or cover gas (liquid-metal-cooled reactors) in the primary coolant system during routine operation and after minor accidents assist in maintaining a low level of radioactive contamination in the coolant, thus providing some reduction of the amount of radioactive materials available for release by a major accident.   However, these systems are neither designed nor located to cope with the large quantities of fission products, activation products, and fuel materials that might be transported into the containment atmosphere by a major accident in the primary system.

B. High-Efficiency Cleanup Systems

Under very favorable conditions, the efficiencies of HEPA filters for the removal of particulate radioactive material and of any one of several impregnated charcoals for the removal of ^{131}I and its gaseous compounds are excellent. The most important requirements that must be met in order to obtain and maintain such efficiencies in actual service are outlined in the following two paragraphs. These requirements are discussed in more detail in (51).

The most critical factors in the adequacy of high-efficiency air-cleaning systems provided for post-accident air cleanup are allocation of enough funds and space for an adequate system, meticulous design, carefully devised and executed operational procedures, constant administrative control, and checks by regulatory authorities to ensure compliance with specifications and regulations. The installed filters and charcoal must be protected from damage and deterioration, particularly during and after an accident, and the structural integrity of the system and all of its components must be maintained. The air must be pretreated by moisture separators and prefilters to minimize the reduction of filter and charcoal efficiency by excessive moisture and clogging by large particles. All components of air-cleaning systems for post-accident cleanup must be designed, constructed, and continually maintained for reliable performance under accident conditions. For example, housings, gaskets, mounting frames, ducts, and dampers must be as reliable as the filter units, charcoal, moisture separators, motors, and fans. In-service testing, inspection, and maintenance of the system and its components must be effective. Any component subject to deterioration or loss of function must be replaced, when necessary, well in advance of the end of its safe service life.

The location of the system has a great effect on both the reliability of high-efficiency filtration-adsorption systems under postulated accident conditions and the ability to test these systems in such a manner that their adequate performance and reliability under accident conditions is assured. Once-through filtration-adsorption systems in secondary containment exhaust lines would function after an accident in an environment very similar to their normal environment. These systems, however, must have far greater structural integrity than that of ordinary air conditioning systems and must be designed and constructed with built-in capabilities for adequate testing, inspection, and maintenance. Component efficiency is less critical in recirculating systems, but they must be rugged, in order to withstand the initial effects of the accident within the containment shell and to perform adequately and reliably thereafter in the hot, high-pressure post-accident containment atmosphere. This atmosphere would also be wet in water-cooled reactor plants and perhaps in CO_2-cooled reactor plants in some accidents.

1. Once-Through High-Efficiency Systems

If the structural integrity of the containment shell is maintained after a major reactor accident, a once-through filter-adsorber system in the secondary containment exhaust line would be required to treat only a relatively small volume of leakage from within the inner containment shell, highly diluted with air from within the secondary containment structure, including air that leaks into the secondary containment from the outside atmosphere. The temperature, pressure, and humidity of the air passing through the system would be essentially ambient, and the loading of radioactive materials would be very low compared with the post-accident loading on a recirculating filter-adsorber system within a containment shell. An air-cleaning system of this type can be routinely tested in its entirety at ambient temperature, pressure, and humidity. Accident conditions would not differ materially from the test conditions, except for the radioactive-material loading. If the system is adequately engineered, constructed, and maintained, and if it is frequently tested, it should prove reliable under accident conditions.

Most types of reactors have once-through high-efficiency systems, provided for post-accident air cleanup, in their secondary (outer) containment structures (see Figures 7-10). Pressurized-water reactors (Figure 6) do not generally have once-through post-accident cleanup systems; exceptions can be found in a few special applications (9,71).

Once-through filter-adsorber systems in the exhaust lines of secondary containment structures are protected by the containment shell and by dilution of the containment air that passes through them. This protection also applies to once-through filter-adsorber systems directly attached to the outside of the shell, if provision is made to dilute the atmosphere from within the containment shell with outside air before it passes through the air-cleaning system. If the containment atmosphere were allowed to pass through the air-cleaning system without dilution, the system would be subjected to an air-stream mixture which, under accident conditions, would be at about the same temperatures, pressures, and humidities as those to which recirculating filter-adsorber systems within the containment shell would be subjected.

Cleanup of air exhausted directly from the containment shell by once-through high-efficiency filter systems has been proposed for several pressurized-water reactors (11,16). For one pair of containment systems (11), the containment air would be cleaned internally by a recirculating system until post-accident pressure had been reduced; the air could then be slowly purged through high-efficiency filters and released to the stack (72).

In the double containment system (16,17) for the Malibu reactor, described in Section II-B-1, air leakage into the annulus between the two pressure-containment shells after a postulated accident would be pumped back into the inner shell to provide maximum containment. If it were necessary to vent the inner shell to the outside atmosphere for reduction of pressure build-up or for post-accident cleanup, air would be slowly purged from within the inner shell through filters, monitored, and exhausted through the stack. However, if the concentration of fission products exceeded permissible limits for release to unrestricted areas, any atmosphere removed from the containment shell would be transferred into special mobile containers for disposal at a remote location (73).

Figure 18 is a schematic diagram of the air-cleaning system for the double containment once proposed for the Ravenswood reactor (74), which is essentially the same design as that for the Malibu reactor. Note the variable-to-large flow under normal conditions and the small flow through the purge system. (Concrete thicknesses in relation to the size of the containment shells have been exaggerated in the diagram in order to show the porous concrete and the valves within the annulus.)

Some of the design characteristics of several once-through high-efficiency air-cleaning systems are given in Table VII and Figure 13. Note that all of the systems in the figure except that for the early (1962 startup) Indian Point 1 reactor (71) include charcoal adsorbers, and those for pressure-suppression plants include high-efficiency moisture separators.

2. Recirculating High-Efficiency Systems

In most pressurized-water reactors, a high-efficiency air-cleaning system is added to the recirculating cooling-coil system provided for cooling the containment atmosphere during normal operations. This air-cleaning system includes moisture separators, prefilters, HEPA filters, and charcoal beds. Because of possible charcoal degradation from contaminants, the charcoal beds (and sometimes other components in the system) are usually bypassed during normal operation. In an emergency, dampers are switched to direct the air flow through the charcoal. These air-cleanup systems must then be able to perform reliably in the post-accident atmosphere within the containment shell, where temperature, pressure, humidity, and particulate concentration would be high. They must be protected from accident-produced pressure surges, shock waves, missiles, and heavy loadings of moisture and particulate materials from the post-accident atmosphere. Prefilters are used to prevent clogging of HEPA filters by heavy particulate loadings. Moisture separators are used to protect HEPA filters from moisture damage and to reduce the amount of moisture that passes through charcoal beds.

Testing, inspection, and maintenance of recirculating systems within the containment shell present special problems. Visual inspection, direct maintenance, and perhaps certain tests can be accomplished only when the reactor is shut down, and remote testing and inspection may necessitate additional or larger penetrations. Also, because moisture separators, HEPA filters, charcoal adsorbers, and other components may become extremely radioactive in operation after a reactor accident, means for removing and

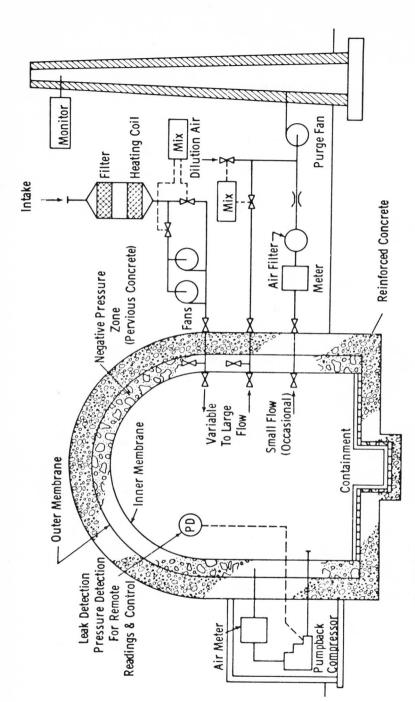

Fig. 18. Schematic diagram of air-cleaning system for proposed double containment concept (74).

TABLE VII

Once-Through Air-Cleaning Systems in the Secondary Containment Exhaust Lines of Pressure Containment Systems for PWR's and Pressure-Suppression Containment Systems for BWR's

	Designer	Electrical power (MW)	Completion date	Design containment shell leakage rate (cfm)	Flow through air-cleaning system (cfm)	Reference
Pressure containment system						
Indian Point 1	Babcock & Wilcox	255	1962			71
Oconee 1 and 2	Babcock & Wilcox	850 each	1971–72	7.1	2000[a]	9
Pressure-suppression containment system						
Oyster Creek	General Electric	640	1967	1		23
Millstone Point	General Electric	650	1969	0.82	2000[a]	22
Dresden 3	General Electric	809	1969	1	2000[a]	24
Browns Ferry 1 and 2	General Electric	1075 each	1970–71	0.97	4000[a]	25

[a] One spare system is provided.

TABLE VIII

Recirculating Air-Cleaning Systems Within the Containment Shells
of Pressure Containment Systems for PWR's

Reactor[a]	Electrical power (MW)	Completion date	Air changes per hour	Parallel systems	Flow per system (cfm)	Reference
Yankee	175	1960	0.77	3	4,000	10
Connecticut Yankee	562	1967	5.37	4	50,000	13
Ginna	470	1969	9.3	4	38,000	75
Indian Point 2	1033	1969	7.4	5	65,000	12
H. B. Robinson 2	~750	1970	11.4	4	100,000	76
Diablo Canyon	1060	1971	7.5	5	65,000	77

[a] Westinghouse Electric Corporation was the designer of these reactors.

handling them remotely must be considered.   Because of the high cost of
ensuring the reliable performance of recirculating filter-adsorber systems
after postulated design-basis accidents, they are more suitable for containment atmosphere cleanup after minor accidents.

Some of the design characteristics of typical recirculating high-efficiency
systems for pressurized-water reactors (10,12,13,75–77) are shown in Table
VIII and Figure 12.   As shown in the figure, all except the system for the
early (1960 startup) Yankee reactor (10) include high-efficiency moisture
separators, most contain prefilters, and charcoal beds are usually bypassed
during normal operating conditions.

C. Chemical-Spray Cleanup Systems

In order to ensure that a recirculating filter-adsorber system will perform
reliably and adequately in the removal of fission products under post-
accident conditions, a system of this type must be protected from pressure
surges, possible missiles, and the high pressures, temperatures, and humidities
that would exist within the containment shell after a postulated loss-of-
coolant accident.   Such protection is very costly.   In an effort to develop
a system that would be less expensive and yet reliable and adequate for post-
accident fission-product removal within the containment shell, much
attention has been directed to examination of the adaptability of containment-atmosphere cooling sprays and pressure-suppression pools for fission
product removal.   Chemical additives to react with iodine and methyl
iodide are under investigation, and the addition of sodium hydroxide or
sodium hydroxide-sodium thiosulfate to the water in the containment-
atmosphere spray cooling system has been proposed for a number of pressurized-water power reactors (12,76–78).

Pressurized-water-reactor containment design has in the past included spray systems for pressure reduction in the unlikely event of a primary steam system rupture.   This enables design of the containment structure for typical values of 60 psig and 290° F.   These spray systems are now also being proposed for use, with slight modification, as devices for the removal of fission products that might be released during the course of a design basis accident.

Current spray systems are designed to inject stored water through spray headers in the upper portion of the containment building.   The concentrated additive, stored outside of the containment, would be mixed with the water so that a specified concentration would be released at the spray nozzles.   The solution would be pumped into the containment through spray headers (the number varies) until the refueling water supply was exhausted; this phase would take approximately $\frac{1}{2}$ hr.   At the end of this period the spray headers would be fed by residual-heat-removal pumps, which would recirculate the water from the main building sump; this switch-over would normally be manual.   Operation of the latter mode could continue for as long as 120 days.   During this recirculation the sump water would also be used by the emergency core cooling systems for shutdown cooling of the reactor core.   The pumps, coolers, and most of the valves are usually located outside the containment shell, where they are less subject to possible accident damage and are more accessible for maintenance and testing. Since they are outside the containment, they must be shielded, and there must be a positive means of containing leakage.   The pumps are normally tested with water through a recirculating loop up to an isolation valve.

The use of chemical sprays requires knowledge of the removal characteristics, thermal and radiation stability, and corrosive nature of the chemical solutions considered.   A research program to study all aspects of reactor containment spray systems was initiated in 1967 by the USAEC with Oak Ridge National Laboratory serving as coordinator for the work (79).   The removal of iodine was initially considered to be of primary importance, and much of the work in the first part of the program was concentrated on this aspect.   Methyl iodide removal and particulate removal are now under investigation as secondary objectives.

When the iodine is removed from the containment atmosphere, it should be converted into a form that is not readily rereleased during the post-accident period.   Both removal and conversion should be accomplished quickly; therefore it is desirable for them to be simultaneous.   Several reagents were suggested initially by industry as logical additives to borated makeup water; these included sodium hydroxide and sodium thiosulfate.   While the major effort in the program has been the examination of these, a number of substitute additives have also been studied.   The effects of parameters such as solution pH, temperature, humidity, etc. are examined using a single drop of solution suspended in a vertical wind tunnel by an upward-flowing gas that contains a regulated amount of the desired contaminant. Large engineering-scale tests of the removal effectiveness of spray solutions under simulated accident conditions are then performed in two facilities: the Nuclear Safety Pilot Plant (NSPP) at The Oak Ridge National Laboratory and the Containment Systems Experiment (CSE) at the Battelle Northwest Laboratory.

The thermal and radiation stability of the spray solutions must also be considered, because the solutions would be subjected to high temperatures (as high as 290° F) and strong radiation fields during the course of an accident. The initial containment environment would contain steam and fission products released during the blowdown; recirculation through the core after the switchover to this mode of operation would expose the solution to the hot fuel element surfaces and radiation field associated with the core.

The possibility that chemical additives in the borated water spray solutions might increase the corrosion of reactor materials is under investigation. As a prerequisite to the use of additives, it must be determined that such use cannot lead to propagation of the accident by violation of the containment or other means. The sodium hydroxide-borate and sodium hydroxide-sodium thiosulfate-borate systems which have been proposed have been found compatible with all but a few materials (86).

The research and development completed to date under both AEC sponsorship and private industry have established that a chemical spray system will be a useful post-accident cleanup system. The removal of molecular iodine, of primary interest in plant siting, is very rapid; the removal of methyl iodide and particulate matter is still under investigation and evaluation. System design requirements, both mechanical and chemical, require careful consideration; however, both the aqueous base-borate solution and the aqueous base-borate-thiosulfate solution appear to be usable. Research on radiolytic hydrogen generation is continuing, and current plant designs consider the use of recombiners to cope with the expected quantities. As a result of the intensive studies of chemical spray systems, a number of recent USAEC research and development reports have been published, including a series issued as an aid in the design of spray systems. These reports (79–90) are recommended for information on the state-of-the-art in the spray research program.

Chemical sprays have been shown to be effective, reliable, and practical for internal post-accident cleanup, and they should be included in the designs of future water-cooled power-reactor plants and perhaps added to existing systems. Practically all pressurized-water reactors under review today are relying on sprays to meet 10 CFR 100 design-basis-accident criteria.

The cleanup potential of pressure-suppression pools is being investigated by both the USAEC and the General Electric Company.

VII. TESTS OF THE EFFICIENCY OF HEPA FILTERS AND IMPREGNATED CHARCOAL BEDS

As mentioned earlier, high efficiency for the filtration of submicron particles and for the adsorption of iodine and its compounds is ensured in nuclear reactor installations by testing new HEPA filters and impregnated charcoal beds before installation, and also by testing them in-place after their installation in air-cleaning systems. The in-place tests are conducted before the system is placed in routine operation and periodically thereafter, and at any time between periodic tests if considered necessary. Several tests are widely used for HEPA filters, and several are used for impregnated charcoal beds, but there is a trend toward standardization in both cases.

A. Standard Efficiency Tests of HEPA Filters

In each of a number of countries in which HEPA or similar filters are widely used, a standard efficiency test or group of tests has been developed for acceptance testing. Methods of measuring the concentration of test particles vary according to the type of particle used, but there is no question that the sensitivity of the detection method must be adequate. The most significant variation among the standard tests is in the types (liquid, solid, or both) and particle size of the aerosols used.

The British standard (91) specifies a polydisperse (i.e., heterogeneous particle size) aerosol of solid sodium chloride crystals (i.e., tiny cubic particles of common salt) with diameters ranging from 0.01 to 1.7 microns and a mass median diameter of about 0.6 micron. The amount of particulate matter penetrating the filter is measured by flame photometry.

The Centre d'Etudes Nucleaires de Fontenay-aux-Roses, which is one of the centers of nuclear studies of France's Commissariat a l'Energie Atomique (CEA), specifies a polydisperse aerosol of uranine (sodium fluorescein, a fluorescent dye) particles with a number mean diameter of 0.15 micron, a mass mean diameter of 0.3 micron, and a range from about 0.085 to 0.5 micron. Aerosol concentrations upstream and downstream of the filter are measured by fluorimetry (92).

The Staubforschungsinstitut des Hauptverbandes der gewerblichen Berufsgenossenschaften (Dust Research Institute) in Bonn, Federal Republic of Germany, uses a combination of tests (93) with solid and liquid particles in three size ranges. The whole filter is tested with a fog of oil droplets of which more than 80% have diameters between 0.3 and 0.5 micron, and the

maximum diameter is less than 1 micron.   Sections of the filter are tested
with radioactively labeled natural atmospheric aerosols ranging in diameter
from several millimicrons to 0.3 micron, with a mean diameter of about 0.07
micron.   Also, sections of the filter are tested with freshly ground quartz
particles of which more than 80 % are between 0.5 and 2 microns in diameter,
the mean diameter is about 1.3 microns, and the maximum is less than 10
microns.   Testing with this range of particle sizes, from a few millimicrons
to several microns, is believed at the Staubforschungsinstitut to be necessary
because the particle size for maximum penetration of the filter medium
depends on the kind and size of particles, the diameters of the fibers in the
filter, the air velocity, and other variables.

Standard HEPA filter acceptance tests in the United States and Canada are,
on the other hand, based on the prediction by the theory of high-efficiency
filtration that a certain particle size, about 0.3 micron, would give maximum
penetration and that particles both larger and smaller than the most
penetrating size would be collected on the filter with higher efficiency (94).
The principal tenet of the theory has been confirmed by extensive experiments; however, experiments and modifications to the theory have indicated
that the particle size for maximum penetration is lower (55) than that
originally predicted and is near 0.07 micron (95).   Filtration research and
development with particles in this size range are continuing in the United
States, as well as in other countries.   However, the 0.3 micron particle diameter, on which the method and equipment for the standard United States test
are based, is considered adequate for acceptance testing of HEPA filters at
the rated air flow for these filters, 5 fpm (2.5 cm/sec) through the filter
medium.

1. Preinstallation Efficiency Testing of HEPA Filter Units with DOP

In the United States the efficiency of new HEPA filters for reactor
installations is determined with thermally generated monodisperse (homogeneous particle size) dioctyl phthalate (DOP) of 0.3-micron particle
diameter, by the manufacturer, following Edgewood Arsenal procedures (96).
In addition, there are two USAEC Quality Assurance Stations (filter testing
facilities) for verifying the efficiency of new filters by the same test procedures.
This service is available on request or when specified by the purchaser (97).

The filter to be tested is installed in a duct; flowing air and the test aerosol
are introduced upstream; and the concentrations of the aerosol in air samples
from the duct upstream and downstream of the filter are measured by
scattered-light photometry.   Efficiency is calculated by dividing the downstream concentration (C_d) by the upstream concentration (C_u), which gives
the penetration (P) as a decimal fraction; multiplying the decimal fraction
by 100 to convert to percentage; and subtracting the percentage penetration

from 100 to give the percentage removed by the filter, which is conventionally referred to as efficiency (E):

$$P = (C_d/C_u) \times 100 \tag{1}$$

$$E = 100 - P \tag{2}$$

The equations above reduce, of course, to

$$E = 100(1 - C_d/C_u). \tag{3}$$

The use of the first two equations rather than the third came about because of practical limitations. Direct measurement of the quantity removed by the filter (Q_r) is not feasible; even if it were, it would probably be far less accurate in practice than measurement of the far smaller quantity that penetrates the filter. Direct measurement of the quantity that penetrates the filter (Q_p) and the quantity introduced upstream (Q_u) are not practical. However, the downstream and upstream concentrations can be measured directly, and their ratio is proportional to the ratio of the quantity that penetrates the filter to the quantity introduced upstream.

The basic relationship is simply $Q_r + Q_p = Q_u$. Thus, if efficiency and penetrations were decimal fractions rather than percentages, they would be related by $E + P = 1$ or $E = 1 - P$. In terms of concentration, penetration would be $P = C_d/C_u$ and efficiency would be $E = 1 - C_d/C_u$.

Each filter is required to meet the specification of at least 99.97 % efficiency for the removal of the standard monodisperse 0.3 micron DOP aerosol. The value 99.97 % is a reasonable compromise between necessary manufacturing tolerance and protection for the user. Most HEPA filters test at substantially higher values (98).

At the filter testing facilities, 500-cfm and larger filter units are tested at two flow rates, the rated flow and 20 % of the rated flow. A significant difference between the results will disclose pinholes and leaks that cannot be found by testing at the rated flow alone or by visual inspection (99). Two-flow testing is based on the pinhole effect (100), which, according to theoretical analyses (101,102), is as follows. Airflow through the intact part of the filter is essentially streamline, and the flow rate is directly proportional to pressure drop. Airflow through a pinhole (or other small defect), however, is turbulent, and the flow rate is proportional to the square root of pressure drop. Therefore, with an increase in overall air flow through the filter and a corresponding increase in pressure drop, proportionately less air passes through the defect. The net effect is an increase in filtration efficiency with increasing flow rate. The pinhole effect would not apply, of course, in cases where the size of the leakage path increased with increasing flow rate. The pinhole effect is well established experimentally (100–104). A statistical analysis of filter test results showed a 95% probability of significant pin-holing if the difference in penetration between the full-flow test and a 20% flow test exceeded 0.01% (105). This is borne out by other work (106,107). It has been recommended (98) that procurement specifications for HEPA filters include a requirement for testing at rated flow and at 20% of rated flow, with a maximum permissible penetration difference of 0.01% between the two flows, for 500-cfm and larger filter units. Testing of smaller units at 20% of rated flow is not considered practical because of questionable results at the low flow velocities.

2. In-Place Testing of HEPA Filter Systems

Experience has shown that a system containing HEPA filters (and/or charcoal adsorbers) should not be considered to have an acceptable efficiency until an in-place test shows that the system meets the efficiency specification. Leakage has been found to be caused by a number of defects in the system, as well as by damaged filters or deterioration of filter units and media. Among the causes of leakage in the system are damaged or improperly seated gaskets; rough or warped frame surfaces; clamps that are too loose, too few in number, or incorrectly positioned; lack of enough frame members; unwelded joints at corners and sides of frames. Some of the causes of leaks in the filters themselves are damage to filter media during installation or other handling; deterioration of filter media in service; and filters that sag because they were installed with their pleats horizontal instead of vertical (103). Recently there has been an increasing awareness of the importance of in-place testing of high-efficiency filter systems. In addition to the initial in-place test, it is very important that each system be tested periodically and, also, after every filter change and whenever there is any reason to suspect that efficiency has decreased. In-place tests of filter systems are proof tests, with a twofold purpose. The total penetration of the system by the test aerosol is measured to establish either that the system has excessive leakage or that it meets the low leakage specification. If leakage is excessive, leaks are located by probing the system, and the defects are plugged or repaired as necessary until the system meets the specification.

A schematic diagram of an in-place test of a HEPA filter system is shown in Figure 19 (51). In the in-place test (103,108,109) used by the USAEC's Oak Ridge National Laboratory (ORNL), polydisperse DOP droplets, produced by atomization with compressed air, are introduced into any convenient air intake that enables thorough mixing with the air stream ahead of the filter bank being tested. The concentration of the unfiltered aerosol is measured on a sample drawn from a point ahead of the filter bank, and that of the aerosol in the filtered air is measured on a sample taken down-stream of the filter bank. Frequently, the downstream sample is taken after

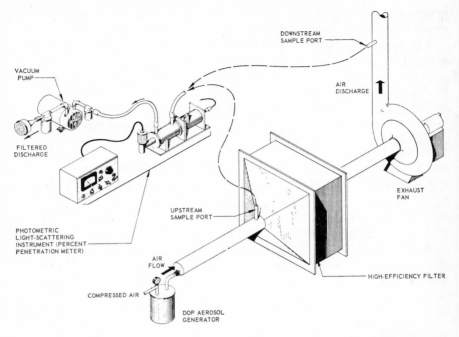

Fig. 19. Schematic of in-place DOP test of HEPA filter system (51).

the exhaust blower to ensure a thoroughly mixed and representative sample. Use of this location is subject to the provision that blower-shaft-seal leakage must be negligible. Sampling lines should not be too long, since loss of sample by deposition within long lines can be severe. The 2 in. thick charcoal beds used in commercial reactor practice are not expected to cause any bias in measurements, however. In in-place DOP tests at ORNL, no difference was found in the measured efficiency of a HEPA filter system when the same system was tested with and without approximately $2\frac{1}{4}$ in. of charcoal (two $1\frac{1}{8}$ in. beds in series) between the HEPA filters and the downstream sampling point (110). In another in-place DOP test, the measured efficiency for a charcoal bed alone was zero.

System efficiency is calculated from the measured concentrations by the same simple equations used for the efficiency of an individual filter unit (Section VII-A-1).

To meet the specification of the in-place test at ORNL, the system must have a minimum efficiency of 99.95% for the removal of the polydisperse DOP droplets. When an in-place test shows an unacceptable efficiency, leakage paths can usually be detected readily by passing the aerosol into the system and probing the downstream side of the bank of filters and the filter mounting frame with a probe connected directly to the photometer.

A proposed standard based on the foregoing test method has been recommended for approval as a USA Standard (111).

In-place testing of future installations can be simplified by careful planning in the design stage. Design of a filter installation should include built-in facilities for conducting in-place testing and access into the duct for sampling instrumentation.

The parallels in method, calculation, and expression of result between tests of new filters and tests of installed systems have caused some ambiguity and misinterpretation. Although the total penetration through the system is a measure of system efficiency, it is not a quantitative measure of the efficiency of the filters themselves, for two reasons. First, the system is usually tested with a compressed-air-generated polydisperse DOP aerosol with a number median diameter of about 0.7 micron, which is not intended as a substitute for the thermally generated monodisperse 0.3 micron DOP specified for testing new filters. Second, the total penetration through the system includes not only penetration through the filters but also penetration through leaks that bypass the filters.

The use of polysize DOP droplets with a number median diameter of 0.7 micron for in-place testing of filter systems, as opposed to the monosize 0.3 micron droplets used for testing filter units, does not adversely affect the detection of leaks or the determination of the efficiency of the system. Most of the particles detected in a system downstream of the high-efficiency filter pass through holes or other continuous leakage paths that are larger by several orders of magnitude than the test particles.

B. Efficiency Tests of Impregnated Charcoal Beds

No standard quality-assurance test of the efficiency of charcoal beds for reactor installations has yet been adopted on a wide scale. However, as discussed below, a standard set of tests will quite possibly be adopted in the near future.

1. Efficiency Testing of New and Recharged Charcoal Beds

Standard testing of the efficiency of new and recharged charcoal beds may include a test for efficiency in removing molecular iodine, a test for efficiency in removing methyl iodide, and perhaps a leak test with a gas other than iodine. Qualification tests will also include a test of the total capacity of representative samples for iodine and methyl iodide. Although a standard set of tests has not yet been widely adopted, these tests have been used for several years.

Standard efficiency tests may be based on the two tests with small amounts of radioactive $^{131}I_2$ and $CH_3^{131}I$ that are used as in-place tests of beds of impregnated charcoal at ORNL (112) (Section VII-B-2) and the extensive tests conducted with $CH_3^{131}I$ at the UKAEA Reactor Development Laboratory at Windscale (65). Alternatively, the tests may be a combination of tests of samples from the beds with $^{131}I_2$ and $CH_3^{131}I$ and a leak test of the entire system with Freon-112*. Freon-112 is a vapor that is sorbable by the charcoal beds and has suitable volatility for leak testing. It can be used for this purpose because the amounts of Freon-112 sorbed by the charcoal during in-place tests do not affect the efficiency of the charcoal for the adsorption of molecular iodine (I_2). If nonradioactive $^{127}I_2$ were used in the necessary quantities for tracer-free leak testing, it might lower the capacity of the charcoal beds for radioactive $^{131}I_2$. The Freon-112 leak test was developed by the Savannah River Laboratory (SRL) and is used routinely at the Savannah River Plant (113,114). If a combination of the Freon-112 leak test and the iodine-131 tests were used, all charcoal beds, whether new or recharged with new charcoal, could be subjected to quality-assurance leak tests with Freon-112, and representative beds and/or representative samples of impregnated charcoal from production runs could be tested for iodine-removal efficiency and capacity with small amounts of radioactive $^{131}I_2$ in a nonradioactive $^{127}I_2$ carrier and with small amounts of $CH_3^{131}I$ in $CH_3^{127}I$.

A set of standards now being written for the USAEC will, if adopted, give the reactor operator a choice of two or more methods (preferences will be indicated) for use at his option, subject to the approval of licensing and regulatory authorities.

Unlike the test of HEPA filters with thermally generated monodisperse 0.3 micron DOP, which is widely accepted as a test of the efficiency of the filters for removal of particulates, the test of charcoal beds with Freon-112 is not considered a test of the efficiency of the beds for the removal of iodine and its gaseous compounds. It is considered a very useful leak test, as is the in-place test of HEPA filters with compressed-air-generated polydisperse DOP.

A number of methods of testing the efficiency of charcoal beds with iodine but without using ^{131}I as a tracer have been used or tried. Among these methods are neutron activation of stable iodine (115), detection of iodine as an oxidant (116), and colorimetric measurement of residual ceric ions after iodine-catalyzed reduction of ceric ions by arsenite (117):

$$2Ce^{4+} + As^{3+} = 2Ce^{3+} + As^{5+}. \tag{4}$$

---

* Trade name of E. I. du Pont de Nemours and Company, Inc. for one of a series of chlorofluorocarbons.

The reaction shown for the last method is actually two reactions: the reduction of ceric ions to cerous ions by iodate ions (which are oxidized to iodine),

$$2Ce^{4+} + 2I^- = 2Ce^{3+} + 2I^0; \tag{5}$$

and the reduction of iodine back to iodate by arsenite ions (which are oxidized to arsenate),

$$2I^0 + As^{3+} = 2I^- + As^{5+} \tag{6}$$

The methods mentioned in the preceding paragraph are not applicable to testing the efficiency of impregnated charcoals used in the United States for removal of methyl iodine and other organic iodides.   This efficiency can only be tested with an organic compound of radioactive iodine which, as a compound, penetrates activated charcoal as easily as methyl iodide does.   The reason is that these impregnated charcoals are specific for the removal of radioactive iodine from methyl iodide by isotopic exchange.   Fortunately, it is both technologically and economically practical to test the efficiency of representative samples of charcoal with the very materials for which charcoal beds are provided, $^{131}I_2$ and $CH_3^{131}I$.

The probable standard efficiency tests outlined in this section would not differ basically from the in-place iodine tests outlined below.   The Freon-112 leak test would differ from in-place tests of used charcoal beds in that the test conditions of air velocity and amount of sorbed water are less restrictive for new beds.

2. In-Place Testing of Charcoal-Adsorber Systems

The charcoal impregnants used in the United States for reactor containment air-cleaning systems are specific for the removal of radioactive iodine from methyl iodide by isotopic exchange.   Therefore, the efficiency of impregnated charcoals in the United States for methyl iodide should be tested by an isotopic-exchange method, preferably with $CH_3^{131}I$ (diluted with $CH_3^{127}I$).   A proof test for leaks with a nonradioactive gas (DuPont Freon-112) has been developed. However, in-place leak testing of charcoal beds in filter-adsorber systems, combined with tests of representative samples of charcoal from the beds with $CH_3^{131}I$ and $^{131}I_2$, should be adequate.   This combination would have the advantage of avoiding testing of the whole system with radioactive material.

At ORNL, charcoal beds are tested in place with small amounts of $^{131}I_2$ and $CH_3^{131}I$ (in $^{127}I_2$ and $CH_3^{127}I$ as carriers) (112).   This method is being adopted in the United States and Canada to some extent, but commercial reactor operators may be reluctant to use it because of the requirements for special precautions and specially trained personnel.   At the Savannah River Plant, charcoal beds are routinely leak tested in place with DuPont Freon-112

(113,114). This method was developed by the Savannah River Laboratory under the sponsorship of the USAEC's Division of Operational Safety. The SRL leak tests are backed up by small-scale laboratory tests of the efficiency and capacity of the activated charcoal for removing molecular iodine by using small amounts of $^{131}I_2$.

VIII. SUMMARY AND CONCLUSIONS

The degree of effectiveness that can be maintained in the removal of air-borne radioactive particulate and chemically reactive materials (the latter primarily iodine-131) from nuclear reactor off-gases during normal operations has been brought to a very high removal efficiency, largely within the past decade. This has been made possible through intensive research and development and thoughtful application of experience by government agencies, designers and operators of reactor plants and other installations in the nuclear field, and manufacturers of air-cleaning components. The most important of these developments are outlined below.

1. High-Efficiency Particulate Air (HEPA) Filters. The glass or glass-asbestos filtering medium in these filters collects and retains at least 99.97 % of the 0.3 micron particles in an air stream introduced into the filter in a standard test. In HEPA filter units the medium is pleated; this provides both moderately high flow capacities (at least 250 cfm per square foot of filter unit face area) and low face velocity (5 fpm) through the filter medium.

2. Improved equipment, methods and procedures for testing HEPA filter units before installation and for testing banks of HEPA filters installed in air-cleaning systems. Each filter unit should be required to meet the USAEC specification of at least 99.97 % efficiency for the removal of standard monodisperse 0.3 micron aerosol particles, or to meet the specifications of the UKAEA, the CEA, the Staubforschungsinstitut, or similar institution. Each installed filter bank should periodically be required to meet the specification of at least 99.95 % efficiency for the removal of standard polydisperse aerosol particles with a number median diameter of about 0.7 micron, or a similar specification.

3. Beds of specially impregnated activated charcoal. These beds are designed to provide moderately high flow rates (about 85 to 200 cfm per square foot of bed surface area) and to prevent or counteract settling that would allow bypass flow of air through leaks. The activated charcoal, which will remove radiactive molecular (elemental) iodine-131 with high efficiency, is specially impregnated so that it will also remove radioactive iodine-131 from methyl iodide (United States) or remove radioactive methyl iodide as such (United Kingdom) with high efficiency.

4. Methods and procedures for testing charcoal (both new and during its service life) and installed banks of charcoal beds that can ensure high efficiency (about 95%) for the removal of radioactive iodine and methyl iodide if relative humidity does not exceed 99% and other conditions are not too severe.

5. High-efficiency moisture separators, which under post-accident conditions would protect HEPA filters from damage, rupture, or loss of efficiency and would protect charcoal beds from flooding and consequent loss of efficiency.

6. Improved methods of design and construction of air-cleaning systems, with emphasis on the ability of air cleaning systems to maintain their leaktightness and structural integrity under any conditions to which they may be subjected and to perform efficiently and reliably under these conditions.

In addition to these recent improvements, some of the most significant contributions to high-efficiency air cleaning at nuclear reactor plants are as follows:

1. Treatment of off-gases in relatively small volumes collected directly from the primary coolant system, during normal operations, before they are diluted with containment air. This makes it possible to delay the passage of the off-gases in holdup volumes for the decay of short-lived radioactive isotopes, after compression if necessary.

2. Use of prefilters to protect and lengthen the service life of HEPA filters.

3. Dilution of treated off-gases with containment ventilation air to further reduce the concentrations of radioactive materials in the effluent gases.

4. Release from a tall stack where necessary.

5. Continuous monitoring of effluent gases, the environment of the reactor plant, and various points within the plant.

6. Thorough testing, inspection, maintenance, administrative control, and regulation.

Separation of radioactive noble gases from reactor off-gases has not been necessary. Holdup in drums or large-diameter pipes has been sufficient for the decay of short-lived noble gas isotopes from water-cooled reactors, and releases of krypton-85 and xenon-133 have been well below the conservative permissible levels (far below in almost all cases). These low levels of release are ensured by gradual, controlled release and continuous monitoring of effluent gases, as well as monitoring of the environment. The release of argon-41 from gas-cooled reactors has also been well below permissible limits and continuously monitored.

Releases of argon-41 and xenon-133 have required only short-term local precautions. However, the projected cumulative contribution of long-lived krypton-85 (half-life about 10 years) to the earth's atmospheric activity in the future is of long-term concern. Research and development efforts to find economically feasible ways of separating krypton (and xenon) from reactor off-gases have been accelerated by the control achieved over other airborne radioactive effluents. The growing emphasis on further reduction of all environmental contamination that could become hazardous to living organisms may soon make removal of radioactive noble gases even more desirable.

There are several methods for the separation of noble gases from other gases. One or more of these methods will undoubtedly become economically feasible. The expense of noble gas separation in nuclear power plants will probably be offset by expenses incurred in increased pollution control by fossil-fuel power plants. In addition, continuing research and development should reduce the cost of noble gas separation. Furthermore, the costs of controlling radioactive airborne effluents will be justified easily if nuclear power is considered an absolute necessity, when fossil fuels become further depleted.

IX. ACKNOWLEDGMENTS

This paper is part of the research sponsored by the U.S. Atomic Energy Commission under contract with the Union Carbide Corporation.

References

1. Operational Accidents and Radiation Exposure Experience Within the USAEC 1943–1964, USAEC Report TID–22268, April 1965.
2. H. B. Piper, "Fact and Fiction Concerning Nuclear Power Safety," *Nuclear News*, **11**(12), 54–59 (1968).
3. All regulations of the United States Government pertaining to atomic energy are published in the U.S. Atomic Energy Commission Rules and Regulations, Title 10 of the *Code of Federal Regulations;* this code is continually updated and revised as necessary.
4. Recommendations of the International Commission on Radiological Protection. *Report of Committee II on Permissible Dose for Internal Radiation*, ICRP Publications 2 (1959), 6 (1962), and 9 (1965), Pergamon Press, New York.
5. Recommendations of the National Committee on Radiation Protection, *Maximum Permissible Body Burdens and Maximum Permissible Concentrations in Air and in Water for Occupational Exposure*, National Bureau of Standards Handbook 69, U.S. Department of Commerce, 1959.
6. J. J. DiNunno et al., *Calculation of Distance Factors for Power and Test Reactor Sites*, USAEC Report TID-14844, March 1962.
7. W. B. Cottrell and A. W. Savolainen, editors, *U.S. Reactor Containment Technology*, USAEC Report ORNL-NSIC-5, 1965.
8. C. G. Lawson, *Emergency Core-Cooling Systems for Light-Water-Cooled Power Reactors*, USAEC Report ORNL-NSIC-24, 1968.

9. Duke Power Company, Oconee Nuclear Station Units 1 and 2, *Preliminary Safety Analysis Report*, Dockets 50-269 and -270.

10. Yankee Atomic Electric Company, *Yankee Nuclear Power Station Final Hazards Summary Report*, Docket 50-29, Vol. 1, June 1962.

11. Florida Light and Power Company, Turkey Point Nuclear Generating Units 3 and 4, *Preliminary Safety Analysis Report*, Dockets 50-250 and 50-251.

12. Consolidated Edison of New York, Inc., Indian Point Nuclear Generating Unit No. 2, *Preliminary Safety Analysis Report*, Docket 50-247.

13. Connecticut Yankee Atomic Power Company, *Facility Description and Safety Analysis for Haddon Neck Plant*, Topical Report No. NYO-3250-5, Docket 50-213.

14. R. F. Denkins and T. E. Northup, "Concrete Containment Structures," *Nuclear Safety*, **6**(2), 194–201, 210–211 (Winter 1964–1965).

15. F. C. Zapp, *Testing of Containment Systems Used with Light-Water-Cooled Power Reactors*, USAEC Report ORNL-NSIC-26, Oak Ridge National Laboratory, August 1968.

16. Department of Water and Power, City of Los Angeles, *Preliminary Hazards Summary Report for Malibu Nuclear Plant Unit 1*, Docket 50-214, November 1963.

17. *Nucleonics*, **21**(2), 17–18 (1963).

18. "Nuclear Safety with Ice Cubes," *Power Engineering*, November 1967. Also see S. J. Weems, W. G. Lyman, and P. B. Haga, "Ice-Condenser Reactor Containment System," *Nuclear Safety*, **11**(3), 215–222 (1970).

19. Indiana and Michigan Electric Company, Cook Nuclear Units 1 and 2, *Preliminary Safety Analysis Report*, Dockets 50-315, 50-316.

20. Tennessee Valley Authority, Sequoyah Nuclear Units 1 and 2, *Preliminary Safety Analysis Report*, Dockets 50-327, 50-328.

21. R. O. Mehann, "N. S. SAVANNAH Operating Experience," *Nuclear Safety*, **4**(4), 126 (1963).

22. The Connecticut Light and Power Company, The Hartford Electric Light Company, and Western Massachusetts Electric Company, *Design and Analysis Report for Millstone Nuclear Power Station*, Docket 50-245.

23. Jersey Central Power and Light Company, *Facility Description and Safety Analysis Report for Oyster Creek Nuclear Power Plant Unit No. 1*, Docket 50-219.

24. Commonwealth Edison Company, The Dresden Nuclear Power Station Unit No. 3, *Preliminary Design and Analysis Report*, Vol. 2, Docket 50-249, Feb. 1966.

25. Tennessee Valley Authority, *Design and Analysis Report for Browns Ferry Nuclear Power Station*, Dockets 50-259 and -260.

26. D. B. Trauger, *Helium-Cooled Nuclear Reactors*, USAEC Report ORNL-TM-2297, October 1968.

27. H. N. Culver, "Containment of Gas-Cooled Power Reactors," *Nuclear Safety*, **4**(4), 90–98, 101–102 (1963).

28. M. Bender, "Safety in Gas-Cooled Power Reactors," *Nuclear Safety*, **4**(4), 21–36, 38 (1963).

29. G. C. Dale, "Control of Airborne Wastes from United Kingdom Gas-Cooled Power Reactors," in *Treatment of Airborne Radioactive Wastes, Proceedings of a Symposium on Operating and Developmental Experience*, International Atomic Energy Agency, New York, August 26–30, 1968, IAEA, Vienna, 1968, pp. 785–792.

30. P. H. W. Wolff, "Design Development of British Gas-Cooled Reactors," *Trans. Am. Nucl. Soc.* **10**(1), 315 (1967).

31. G. Brown and J. D. Thorn, "The Gas-Cooled Reactor," *Trans. Am. Nucl. Soc.* **10**(1), 315–316 (1967).

32. A. L. Habush and A. M. Harris, "Fort Saint Vrain High-Temperature Gas-Cooled Reactor," *Trans. Am. Nucl. Soc.*, **10**(1), 320–321 (1967).

33. D. B. Trauger, Oak Ridge National Laboratory, personal communication to G. W. Keilholtz, Nov. 11, 1969.

34. G. W. Keilholtz and G. C. Battle, Jr., "Fission-Product Release and Transport in Liquid-Metal-Cooled Fast Breeder Reactors," *Nuclear Safety*, **9**(6), 494–509 (1968).

35. G. W. Parker and C. J. Barton, "Fission Product Release," *Technology of Nuclear Reactor Safety*, Vol. II (to be published by the Massachusetts Institute of Technology Press) Chapter 18.

36. G. W. Keilholtz and G. C. Battle, Jr., *Fission Product Release and Transport in Liquid Metal Fast Breeder Reactors*, USAEC Report ORNL-NSIC-37, March 1969.

37. D. G. Jacobs, *Sources of Tritium and Its Behavior Upon Release to the Environment*, USAEC Report TID-24635, 1968.

38. J. O. Blomeke and F. E. Harrington, *Management of Radioactive Wastes at Nuclear Power Stations*, USAEC Report ORNL-4070, January 1968.

39. J. Mishima, "Methyl Iodide Behavior in Systems Containing Airborne Radio. iodine," *Nuclear Safety*, **9**(1), 35 (1968).

40. J. Mishima, *Review of Methyl Iodide Behavior in Systems Containing Airborne Radioiodine*, USAEC Report BNWL-319, Pacific Northwest Laboratory, June 1966.

41. W. B. Cottrell, *Nuclear Safety Program Annual Progress Report for Period Ending December* 31, 1966, USAEC Report ORNL-4071, Oak Ridge National Laboratory, March 1967, pp. xxii and 127.

42. G. W. Keilholtz, *Filters, Sorbents, and Air Cleaning Systems as Engineered Safeguards in Nuclear Installations*, USAEC Report ORNL-NSIC-13, Oak Ridge National Laboratory, October 1966.

43. R. H. Barnes et al., *Studies of Methyl Iodide Formation Under Nuclear Reactor-Accident Conditions*, USAEC Report BMI-1829, Battelle Memorial Institute February 1968.

44. G. W. Keilholtz, "Removal of Radioactive Noble Gases from Off-Gas Streams," *Nuclear Safety*, **8**(2), 155–160 (Winter 1966–1967).

45. J. H. MacMillan, The Babcock and Wilcox Company, unpublished information, December 2, 1958.

46. Air Reduction Company, Inc., *Noble Gas Recovery Study—Maritime Nuclear Ship SAVANNAH*, Final Report, April 5, 1965 (undocumented).

47. J. M. Holmes, Oak Ridge National Laboratory, unpublished information, October 9, 1963.

48. S. Blumkin et al., "Preliminary Results of Diffusion Membrane Studies for the Separation of Noble Gases from Reactor Accident Atmospheres," Paper presented at the Ninth AEC Air Cleaning Conference, Boston, Mass., September 13–16, 1966.

49. R. H. Rainey, "Separation of Noble Gases from Air by Permselective Membranes," in *Nuclear Safety Program Annual Progress Report December* 31, 1967, USAEC Report ORNL-4228, Oak Ridge National Laboratory, pp. 173–182.

50. J. R. Merriman, J. H. Pashley, and S. H. Smiley, *Engineering Development of an Absorption Process for the Concentration and Collection of Krypton and Xenon, Summary of Progress Through July* 1, 1967, USAEC Report K-1725, Oak Ridge Gaseous Diffusion Plant, December 1967.

51. C. A. Burchsted and A. B. Fuller, *Design, Construction and Testing of High-Efficiency Air Filtration Systems for Nuclear Application*, USAEC Report ORNL-NSIC-65, January 1970.

52. W. S. Durant et al., *Activity Confinement System of the Savannah River Plant Reactors*, USAEC Report DP-1071, Savannah River Laboratory, August 1966.

53. A. H. Peters, *Application of Moisture Separators and Particulate Filters in Reactor Containment*, USAEC Report DP-812, Savannah River Laboratory, December 1962.

54. R. D. Rivers and J. L. Trinkle, *Moisture Separator Study*, USAEC Report NYO-3250-6, Connecticut Yankee Atomic Power Company, June 1966. (These tests were conducted by the Clean Air Group Research Laboratory, American Air Filter Company.)

55. M. W. First, "Filters, Prefilters, High Capacity Filters, and High Efficiency Filters; Review and Projection," *Proceedings of the Tenth AEC Air Cleaning Conference*, New York, August 28, 1968, USAEC Report CONF-680821, December 1968, pp. 65–78.

56. C. A. Burchsted, comment in "Discussion," *Treatment of Airborne Radioactive Wastes, Proceedings of a Symposium on Operating and Developmental Experience*, International Atomic Energy Agency, New York, August 26–30, 1968, IAEA, Vienna, 1968, p. 191.

57. J. Mishima, "Methyl Iodide Behavior in Systems Containing Airborne Radioiodine," *Nuclear Safety*, **9**(1), 35–42 (1968), esp. p. 40.

58. W. E. Browning, Jr., comment in "Discussion," *Proceedings of Ninth AEC Air Cleaning Conference*, Boston, Mass., Sept. 13–16, 1966, USAEC Report CONF-660904, Vol. 2, January 1967, pp. 1175–1193.

59. R. E. Adams, W. E. Browning, Jr., W. B. Cottrell, and G. W. Parker, *The Release and Adsorption of Methyl Iodide in the HFIR Maximum Credible Accident*, USAEC Report ORNL-TM-1291, Oak Ridge National Laboratory, October 1965.

60. R. D. Ackley and R. E. Adams, *Removal of Radioactive Methyl Iodide from Steam-Air Systems (Test Series II)*, USAEC Report ORNL-4180, Oak Ridge National Laboratory, October 1967.

61. *Safety Evaluation by the Division of Reactor Licensing*, USAEC, in the Matter of Consolidated Edison Company of New York, Inc., Indian Point Nuclear Generating Unit No. 2, Peekskill, New York, Docket No. 50-247, Aug. 25, 1966, p. 64.

62. *Safety Evaluation by the Division of Reactor Licensing*, USAEC, in the Matter of Philadelphia Electric Company, Peach Bottom Atomic Power Station Unit Nos. 2 and 3, Peach Bottom Township, York County, Pennsylvania, Docket Nos. 50-277, 50-278, November 7, 1964, p. 34.

63. R. D. Ackley and R. E. Adams, *Trapping of Radioactive Methyl Iodide from Flowing Steam-Air; Westinghouse Test Series*, USAEC Report ORNL-TM-2728 (December 1969).

64. R. E. Adams, R. D. Ackley, and Z. Combs, "Trapping of Radioactive Methyl Iodide from Flowing Steam-Air: Westinghouse Test Series," in *Nuclear Safety Program Annual Progress Report December* 31, 1969, USAEC Report ORNL-4511, March 1970, pp. 49–50 (Section 3.3).

65. D. A. Collins, L. R. Taylor, and R. Taylor, *The Development of Impregnated Charcoals for Trapping Methyl Iodide at High Humidity*, UKAEA Report TRG-1300(W), 1967.

66. R. E. Adams, R. D. Ackley, and Z. Combs, "Weathering of Impregnated Charcoals Used for Trapping Radioiodine," in *Nuclear Safety Program Annual Progress Report December* 31, 1969, USAEC Report ORNL-4511, March 1970, pp. 50–51 (Section 3.4).

67. M. I. Goldman, "United States Experience in Management of Gaseous Wastes from Nuclear Power Stations," *Treatment of Airborne Radioactive Wastes, Proceedings*

of a Symposium on Operating and Developmental Experience, International Atomic Energy Agency, New York, New York, August 26–30, 1968, IAEA, Vienna, 1968, pp. 763–783.

68. "Voluminous Record on Power Plant Environmental Effects," *Nuclear Industry*, **16**(1), 27–44 (1969).

69. *Hearings before the Joint Committee on Atomic Energy*, Congress of the United States, Ninety-First Congress, First Session on Environmental Effects of Producing Electric Power, October 28–31 and November 4–7, 1969, Part I, U.S. Govt. Printing Office, Washington, D.C.

70. F. J. Viles, Jr., comment in "Discussion," *Proceedings of the Tenth AEC Air Cleaning Conference*, New York, August 28, 1968, USAEC Report CONF-680821, December 1968, pp. 53–54.

71. *Report on Hazards Analysis and Design for Containment Vessel of Consolidated Edison Thorium Reactor*, Exhibit K-4, Docket 50-3, Appendix A, August 1958.

72. G. W. Keilholtz, "Air Cleaning Systems as Engineered Safeguards in Nuclear Reactor Containment," *Nuclear Safety*, **8**(4), 360–370 (1967).

73. H. B. Piper, "Descriptions of Specific Containment Systems," *U.S. Reactor Containment Technology*, Wm. B. Cottrell and A. W. Savolainen, (Eds.), USAEC Report ORNL-NSIC-5, Oak Ridge National Laboratory, August 1965, p. 7.134 (Chapter 7).

74. *Preliminary Hazards Summary Report of Consolidated Edison Company of New York, Inc.*, Ravenswood Nuclear Generating Unit A, 1963, USAEC Report NP-12467, Technical Information Service Extension, 1962.

75. Rochester Gas and Electric Corporation, *Preliminary Facility Description and Safety Analysis Report for Brookwood Nuclear Station Unit No. 1*, Docket 50-244.

76. Carolina Power and Light Company, *Preliminary Facility Description and Safety Analysis Report for H. B. Robinson Unit 2*, Docket 50-261.

77. Pacific Gas and Electric Company, *Preliminary Safety Analysis Report for Nuclear Plant, Diablo Canyon Site*, Docket 50-275.

78. Consumers Power Company, *Facility Description and Safety Analysis Report for Palisades Plant*, Docket 50-255.

79. T. H. Row, Coordinator, *Spray and Pool Absorption Technology Program*, USAEC Report ORNL-4360, April 1969.

80. T. H. Row, "Reactor Containment Spray Program," *International Atomic Energy Agency Symposium on Operating and Developmental Experience in the Treatment of Airborne Radioactive Wastes*, SM-110/26, August 1968, New York.

81. L. F. Parsly and J. K. Frangst, *Removal of Iodine Vapor from Air and Steam-Air Atmospheres in the Nuclear Safety Pilot Plant*, USAEC Report ORNL-4253, 1968.

82. L. F. Parsly, Pilot Plant Studies of Methyl Iodide Cleanup by Sprays. *Nuclear Applications and Technology*, **8**(1), 13–22 (1970).

83. L. F. Coleman, R. K. Hilliard, J. D. McCormack, *Nuclear Safety Quarterly Report February–April*, 1969, BNWL-1084, Pacific Northwest Laboratory, Richland, Wash., June 1969, pp. 2.1–2.4.

84. T. H. Row, L. F. Parsly, and H. E. Zittel, *Design Considerations of Reactor Containment Spray Systems—Part I*, USAEC Report ORNL-TM-2412, Part I, April 1969.

85. C. S. Patterson and W. T. Humphries, "Removal of Iodine and Methyl Iodide from Air by Liquid Solutions," *Design Considerations of Reactor Containment Spray Systems—Part II*, USAEC Report ORNL-TM-2412, Part II, August 1969.

86. J. C. Griess and A. L. Bacarella, "The Corrosion of Materials in Spray Solutions,"

Design Considerations of Reactor Containment Spray Systems—Part III, USAEC Report ORNL-TM-2412, Part III.

87. L. F. Parsly, "Calculation of the Partition Coefficient for Iodine Between Water and Air," *Design Considerations of Reactor Containment Spray Systems—Part IV*, USAEC Report ORNL-TM-2412, Part IV, in press.

88. J. C. Griess, T. H. Row, and C. D. Watson, "Protective Coatings Tests," *Design Considerations of Reactor Containment Spray Systems—Part V*, USAEC Report ORNL-TM-2412, Part V.

89. L. F. Parsly, "The Heating of Spray Drops in Air-Steam Atmospheres," *Design Considerations of Reactor Containment Spray Systems—Part VI*, USAEC Report ORNL-TM-2412, Part VI.

90. L. F. Parsly, "A Method for Calculating Iodine Removal by Sprays," *Design Considerations of Reactor Containment Spray Systems—Part VII*, USAEC Report ORNL-TM-2412, Part VII.

91. *British Standard Method of Test for Low-Penetration Air Filters (other than for Air Supply to I.C. Engines and Compressors)*, B.S. 3928: 1965, British Standards Institution, London, 1965.

92. J. Pradel and J. Brion, "Methode Sensible de Mesure de l'Efficacite des Filtres a Haute Efficacite au Moyen d'un Aerosol d'Uranine," *Treatment of Radioactive Wastes, Proceedings of a Symposium on Operating and Developmental Experience*, International Atomic Energy Agency, New York, August 26–30, 1968, IAEA, Vienna, 1968, pp. 279–289.

93. D. Hasenclever, "The Testing of High-Efficiency Filters for the Collection of Suspended Particles," *International Symposium on Fission Product Release and Transport Under Accident Conditions*, Oak Ridge, Tennessee, April 5–7, 1965, USAEC Report CONF-650407, pp. 805–813.

94. Irving Langmuir, *Theory of Filtration of Smokes*, OSRD Report No. 865, Office of Technical Services, Washington, D.C., 1942.

95. W. L. Anderson, comments in "Discussion," *Proceedings of the Tenth AEC Air Cleaning Conference*, New York, August 28, 1968, USAEC Report CONF-680821, December 1968, pp. 145–146.

96. Quality Assurance Directorate, U.S. Army Edgewood Arsenal, Instruction Manual for Q76 DOP Filter Testing Penetrometer, Document No. 136-300-195A, and Instruction Manual for Q107 DOP Filter Testing Penetrometer, Document No. 136-300-175A, Edgewood Arsenal, Maryland. (These manuals are replacing MIL-STD-282 and will eventually be replaced by a USA standard.)

97. *USAEC Health and Safety Information Bulletin on Filter Unit Inspection and Testing Service*, issued annually by the Division of Operational Safety, USAEC, Washington, D.C.; see also USAEC Health and Safety Information Issue No. 212, *Minimal Specification for the Fire-Resistant High-Efficiency Filter Unit*, June 25, 1965.

98. C. A. Burchsted, "Requirements for Fire-Resistant High-Efficiency Particulate Air Filters," *Proceedings of Ninth AEC Air Cleaning Conference*, Boston, Mass., Sept. 13–16, 1966, USAEC Report CONF-660904, Vol. 1, January 1967, pp. 62–74.

99. Humphrey Gilbert, "Octennial History of the Development and Quality of High-Efficiency Filters for the U.S. Atomic Energy Program," *Treatment of Airborne Radioactive Wastes, Proceedings of a Symposium on Operating and Developmental Experience*, International Atomic Energy Agency, New York, N.Y., August 26–30, 1968, IAEA, Vienna, 1968, pp. 227–234.

100. H. W. Knudsen and L. White, *Development of Smoke Penetration Meters*, Report NRL-0-2642, U.S. Naval Research Laboratory, Sept. 14, 1945.

101. J. W. Thomas, "Aerosol Penetration Through Pinholed Filters," *Health Phys.*, **11**, 667–673 (1965).

102. F. E. Adley and D. E. Anderson, "The Effects of Holes on the Performance Characteristics of High-Efficiency Filters," *Proceedings of Eighth AEC Air Cleaning Conference*, Oak Ridge National Laboratory, Oct. 22–25, 1963, USAEC Report TID-7677, March 1964, pp. 494–507.

103. E. C. Parrish and R. W. Schneider, *Tests of High Efficiency Filters and Filter Installations at ORNL*, USAEC Report ORNL-3442, Oak Ridge National Laboratory, May 17, 1963.

104. Earl Stafford and W. J. Smith, *Ind. Eng. Chem.*, **43**, 1345 (1951).

105. C. A. Burchsted, Oak Ridge National Laboratory, unpublished data, July 1966.

106. F. E. Adley, Hanford Occupational Health Foundation, personal communication to C. A. Burchsted, Oak Ridge National Laboratory, 1966.

107. R. H. Knuth, "Performance of Defective High-Efficiency Filters," *Ind. Hyg. Assoc. J.*, **26**, 593–600 (November–December 1965).

108. R. W. Schneider, "In-Place Testing of High-Efficiency Filters," *Nucl. Safety*, **4**(3), 56–58 (1963).

109. W. W. Goshorn and A. B. Fuller, "Particulate Filter Testing and Inspection Program," *Nucl. Safety*, **2**(2), 37–38 (1960).

110. E. C. Parrish, Oak Ridge National Laboratory, personal communication, March 1968.

111. Efficiency Testing of Air-Cleaning Systems Containing Devices for Removal of Particulates, Proposed USA Standard, United States of America Standards Institute Task Group N5.2.11, August 1966.

112. J. H. Swanks, "In-Place Iodine Filter Testing," *Proceedings of Ninth AEC Air Cleaning Conference*, Boston, Mass., Sept. 13–16, 1966, USAEC Report CONF-660904, Vol. 2, January 1967, pp. 1092–1104.

113. W. S. Durant et al., *Activity Confinement System of the Savannah River Plant Reactors*, USAEC Report DP-1071, Savannah River Plant, August 1966.

114. D. R. Muhlbaier, *Standardized Nondestructive Test of Carbon Beds for Reactor Confinement Applications, Final Progress Report, February to June 1966*, USAEC Report DP-1082, Savannah River Plant, July 1967.

115. C. A. Gukeisen and K. L. Malaby, "In-Place Testing of Charcoal Filter Banks at Ames Laboratory Research Center (ALRR)," *Proceedings of Ninth AEC Air Cleaning Conference*, Boston, Mass., Sept. 13–16, 1966, USAEC Report CONF-660904, Vol. 2, January 1967, pp. 1063–1068.

116. W. G. Thomson and R. E. Grossman, "In-Place Testing for Iodine Removal Efficiency Using an Electronic Detector," *Proceedings of Ninth AEC Air Cleaning Conference*, Boston, Mass., Sept. 13–16, 1966, USAEC Report CONF-660904, Vol. 2, January 1967, pp. 1134–1149.

117. F. J. Viles, Jr. and L. Silverman, "In-Place Iodine Removal Efficiency Test," *Proceedings of Ninth AEC Air Cleaning Conference*, Boston, Mass., Sept. 13–16, 1966, USAEC Report CONF-660904, Vol. 2, January 1967, pp. 1108–1132.

Thermal Deposition of Aerosols

JAMES A. GIESEKE

Battelle
Columbus Laboratories
Columbus, Ohio

I. INTRODUCTION

An aerosol particle suspended in a gas in which there exists a temperature gradient experiences a force directed toward the cooler temperatures. The magnitude of this thermal force is dependent on the physical properties and characteristics of the particle and the surrounding gas, as well as the magnitude of the temperature gradient. The thermal force can be considered as one type of radiometric force and defined as the result of energy transfer between the particle and the surrounding gas through the thermal motion of the gas molecules (1). Thermal deposition of airborne particles occurs when the thermal-force-induced movement, or thermophoresis, of particles through the suspending gas causes them to be deposited on a surface which is cooler than the gas.

Thermal deposition can be a naturally occurring nuisance, as when it contributes to the rapid soiling of cool windows and room walls, or it can be exploited as a means of removing particles from an air stream. Although thermal precipitation has been used extensively as a method of sampling airborne particles for observation or analysis, it has been used to only a limited extent as a method for cleaning effluent air streams in air pollution control procedures. Interestingly, the use of thermal precipitation as an air cleaning technique was suggested by Aitken (2) as early as 1884. Perhaps the most promising applications of thermal precipitation in air cleaning are possible when the dusty gas is hot and can be cooled, or when thermal deposition can be added as a contributing mechanism in a device whose effectiveness is based on other collection mechanisms.

The discussions in this chapter are directed toward identifying those general occurrences and processes which result from thermal deposition of particles; reviewing the nature, prediction, and measurements of thermal forces; reviewing their current applications; and reviewing possible areas of application in air cleaning processes.

II. BACKGROUND

Thermal precipitation of aerosol particles is readily observable in everyday surroundings. The uneven soiling or deposition of dirt on indoor walls of heated rooms has long been a problem, and thermal forces have been identified as a major cause of this deposition. In 1884, Aitken (2) and Lodge (3) reported that when a room is heated by direct radiation (as by an open fire) less dirt is deposited on the walls than when a room is heated by convection. The reason given is that the walls of a radiation-heated room remain warmer than the air. Gibbs (4) and Poynting and Thomson (5) also postulated that thermal forces contribute to dust deposition on room walls and offered as evidence the fact that dust deposits are concentrated above radiators and lamps and on the cooler portions of the ceiling between the laths and rafters.

Experimental studies of the causes of dirt patterns on walls were performed by Hooper (6) and Bonnell and Burridge (7). The major conclusion from both studies was that thermal deposition of airborne dust particles is the cause of pattern staining of walls. Nielsen (8) found that the temperature gradients measured near room walls were great enough to cause significant thermal forces on particles as far as 2 mm from the wall and he concluded that these thermal forces accounted for the pattern staining observed.

Clusius (9), Bryhni (10), and Roots and Walker (11) recognized that thermal forces are important as a cause of wall deposition, especially near heaters, and sought procedures for minimizing this effect. It is to be expected that thermal precipitation will be greatest near a heater where the effluent warm air passes over a cooler wall giving rise to large temperature gradients. The procedures recommended by Clusius (9) and Bryhni (10) for reducing thermal deposition at these locations are to direct the heated air so that it mixes with room air away from wall surfaces, to maintain the wall surface at higher temperatures, or to use heating surfaces of lower temperature so that the heating air is not at such a high temperature. Each of these procedures is intended to reduce the temperature gradient at the wall and subsequently reduce thermal deposition. Roots and Walker (11) report that the design of thermostats for temperature control also affects dirt deposition on walls.

Thermal deposition can be of significance in other practical situations. Perhaps of most importance is the deposition of dust particles from air streams onto heat-transfer surfaces. It is to be expected that thermal precipitation will lead to higher fouling rates for heat-transfer surfaces when air is cooled than when air is heated. However, to the knowledge of the author, no allowance is routinely made in fouling rates for thermal deposition, nor has any experimental study been performed to quantitatively assess the effect that thermal precipitation may have in increasing fouling rates. However, particle deposition onto cooling surfaces has been recognized to be the result of thermophoresis and awareness of these effects is evidenced in some rather diverse heat-transfer situations. For example, Drake and Harnett (12) state that thermal deposition will certainly play a part in the overall deposition of carbon particles on heat-transfer surfaces in the oil-fired superheaters of power generation boilers. In addition, the data presented by Drake and Harnett also show that vanadium pentoxide particles deposit at a faster rate on the boiler-tube surfaces when the tube surface temperatures are lower indicating that thermal precipitation may have contributed to the deposition rate.

Hawes and Garton (13) reviewed information regarding heat exchanger fouling by dust suspensions. Most of the data were from studies related to catalyst circulation and nuclear reactor coolants. The feasibility of using suspended particulate matter in gaseous coolants for reactors has been investigated because of the enhanced cooling expected from increasing the heat capacity of the coolant and increasing the heat-transfer coefficient. Boothroyd (14) summarizes the findings of numerous experiments which show heat-transfer rates to dust suspensions to be greater than heat-transfer rates to clean gases. However, the increase is also shown to be greater in the case of heating rather than for cooling since surfaces become fouled by dust particles. It has been repeatedly observed that when a dust suspension is cooled, particles deposit on the cool heat-exchange surfaces; whereas, if the suspension is heated, no deposit is formed. The observation that deposition is predominantly on the cooler wall surfaces has led experimenters to the conclusion that thermal deposition is the controlling factor (13–17). The problems of particle buildup can be reduced considerably by proper choice of flow conditions and particle size. Higher flow rates and larger particles reduce the problem of deposit formation primarily because of reentrainment of particles from the surface.

When particles are sampled from a hot gas, thermal precipitation can contribute to particle loss in the sampling conduits by enhancing deposition on the conduit walls. Particle loss by various mechanisms in sampling lines was postulated by Postma and Schwendiman (18) and they performed analyses to estimate the magnitudes of such losses (18,19). For the cases they considered, particle loss by thermal deposition was estimated as low.

Although wall discoloration and the practical applications of thermal forces have resulted from enhanced particle deposition on surfaces, thermophoresis was first observed as the inhibition of dust deposition and the formation of a dust-free space near a heated surface. In 1870 Tyndall (20) reported that dust particles in a gas were repulsed by a heated surface held in the dusty gas. Subsequent studies by Rayleigh (21,22), Aitken (2), Lodge (3), and Lodge and Clark (23) verified the experimental observations of particle movement away from hot surfaces and also identified the tendency of cold surfaces to collect particles. The dust-free space was found to extend out from the heated surface to an observable boundary as shown in

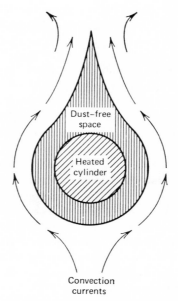

Fig. 1. Dust-free space around a heated cylinder.

Figure 1. Dust particles suspended in a gas were found to move with convection currents but not to penetrate the dust-free space. Lodge and Clark (23) reported experimental results which showed that the width of the dust-free space increased with increasing temperature of the heated body and decreased with increasing gas pressure and decreasing molecular weight of the gas.

Additional experimental verification of the effects of temperature and gas pressure was reported by Watson (24). Empirical relationships were developed by Miyake (25) and Watson (24) which give the width of the dust-free space as dependent on gas pressure, convective heat loss, and heated-body temperature and geometry. Analytical studies of the dust-free space near hot vertical plates and horizontal cylinders were made by Zernik (26). He based his analyses on combinations of aerodynamic and thermal forces and employed the theory of Epstein (27) for the thermal force. Comparisons of the predicted width of the dust-free space with the measurements of Watson (24) show good agreement (26) for horizontal cylinders. This agreement may be fortuitous since the experimental Grashof numbers were outside the range for which the theory was derived and would be expected to be valid.

Quantitative measurements of the velocity of particle motion in a temperature gradient were first attempted by Cawood (28). The experimental technique was to measure the time required for the formation of the dust-free space after it had been destroyed by a blast of turbulent air. Uncertainties about the magnitude of the temperature gradient limit the usefulness of these data.

A different technique was employed by Paranjpe (29) with more quantitative success. A temperature gradient was established between two enclosed horizontal plates. A puff of smoke was then introduced between the plates, and the velocity at which the upper boundary of the smoke cloud moved toward the cooler, lower plate was measured. Since first used by Paranjpe, the basic experimental concept of horizontal plates, with a heated upper plate, has been employed in most experimental techniques.

The next basic experimental advance was made by Rosenblatt and LaMer (30) who observed the motion of single particles suspended in a controlled, vertical temperature gradient between two enclosed horizontal plates. Nearly all experimental studies since that time (31–35) have used similar equipment with only minor modifications of the design used by Rosenblatt and LaMer. Exceptions have included the use of specially designed flowing-gas systems by Derjaguin, Storozhilova, and Rabinovich (36), of a conventional thermal precipitator by Schadt and Cadle (37), and of thermal precipitators with radial gas flow by Orr and Wilson (38) and Keng and Orr (39).

The practical application of thermal precipitation was first suggested by Aitken (2) who designed small collectors with both parallel-plate and concentric-cylinder geometries.

In view of the emphasis placed on reducing air pollutant emissions in modern times, it is of historical interest to note that as early as 1884 Aitken (2) suggested, ". . . perhaps the application (of thermal precipitation) of most general interest would be towards the prevention of smoke, or rather the prevention of the escape of smoke into the atmosphere. Whatever interest, however, it may have in this way, it is clear it can never meet with general adoption, save under compulsion, as it will effect no saving in fuel, such as would result from more perfect forms of combustion." He further suggested that an air cleaner that destroyed the buoyancy of effluent-gas plumes might require that the gases be reheated before release. Bancroft (40), using concentric cylinders, found that thermal precipitation was too slow to be useful for his purposes. Since that time, Blacktin (41) experimented with equipment capable of handling larger air flows and a later patent was issued to Sherwood (42) for a residential air cleaner employing thermal precipitation.

The most widespread use of thermal precipitation has been in aerosol samplers. These are most often used when high efficiencies are required for small particles, when sampling rates can be low, and when the collected particles are to be microscopically examined. The first thermal precipitator widely used for sampling aerosol particles was designed by Green and Watson (43,24) and followed the technique of Aitken (2) who placed a heated wire between two parallel, vertical glass plates to cause deposition of particles on the plates. The original design of Green and Watson is still used extensively although modifications of this design and various other precipitator designs are also widely used.

III. THERMAL FORCES

The rate of thermal precipitation of aerosol particles is dependent on the thermal force exerted on the particles, the geometry of the precipitator, and the characteristics of the gas flow through the precipitator. Consideration of thermal forces is a necessary first step in an analysis of thermal precipitation. The modern theories and experimental studies of thermophoresis have been carefully reviewed by Waldmann and Schmitt (44); earlier studies were discussed by Rosenblatt and LaMer (30).

The thermal force must be treated according to the size of the aerosol particle in relation to the mean free path of the gas molecules. The free-molecule (or small-particle) regime is characterized by values for the ratio of mean free path (λ) to particle radius (r) which are large compared to unity (i.e., $\lambda/r \gg 1$). Similarly, the continuum (or large-particle) regime and the intermediate or transition regime are characterized by $\lambda/r \ll 1$ and $\lambda/r \approx 1$, respectively. In the following discussion of thermal forces, each of these three regimes will be considered individually.

A. Theory for Small Particles

The first reasonably accurate prediction of thermal forces on aerosol particles was a simple kinetic-theory model by Einstein (45) developed to predict forces on radiometer vanes. Similarly qualitative, kinetic-theory models were developed by Cawood (28), Clusius (46), and Stetter (47). A more recent theory developed independently by Derjaguin and Bakanov (48) and by Waldmann (49) gives quite accurate thermal-force predictions for small particles and is based on momentum exchange between the impinging gas molecules and the particle. Another recent theory developed by Mason and Chapman (50) is based on a kinetic-theory approach using a combination of elastically diffuse and specular gas-molecule reflections from the particle surface. The thermal-force predictions with this theory are not substantiated by experiment. More recently, Monchick, Yun, and Mason (51) have used a kinetic-theory approach for the inelastic collision case and have found agreement with Waldmann's theory. They have also reported that the assumption of inelastic or elastic collisions has a significant effect on the predicted thermal force.

The force on a small particle according to the Bakanov-Derjaguin and Waldmann theories (44) is given by

$$\mathbf{F}^* = -\frac{32}{15}\frac{r^2 k_{tr}}{\bar{c}}\nabla T \tag{1}$$

where r is the particle radius; k_{tr} is the translational contribution to the thermal conductivity of the gas which for polyatomic gases is given by $k_{tr} = 15K\eta/4m$ where K is the Boltzmann constant, η is the absolute gas viscosity, and m is the molecular mass for the gas; ∇T is the temperature gradient; and $\bar{c}$ is the mean thermal velocity of the gas molecules and is given by

$$\bar{c} = \sqrt{\frac{8KT}{\pi m}} \tag{2}$$

where T is the absolute gas temperature. This force equation is based on momentum transfer from gas molecules impinging and rebounding from the particle surface. The velocity distribution of the impinging molecules is assumed to be undisturbed by the presence of the particle and the rebounding molecules are divided into a fraction $(1 - a)$ which is reflected specularly while maintaining the same molecular velocity distribution, and a fraction, a, which is accommodated thermally with the particle surface and then reemitted diffusely.

The frictional resistance to a particle moving with velocity, $\mathbf{v}$, through a gas at the same temperature as the particle has been found by Epstein (52) and can be written (44) as

$$\mathbf{F_D} = -\frac{32r^2p}{3}\left(1 + \frac{\pi}{8}a\right)\mathbf{v} \tag{3}$$

where p is the gas pressure. The case when the particle is at a different temperature than the gas has been considered by Slinn, Shen, and Mazo (53). For constant-velocity movement of a highly conductive particle due to the thermal force, the thermophoretic velocity is found from eqs. 1 and 3 to be

$$\mathbf{v} = -\frac{8k_{tr}}{5(8 + \pi a)p}\,\nabla T \tag{4}$$

It is evident that for small particles, or dilute gases such that $\lambda/r \gg 1$, the thermophoretic velocity of aerosol particles is independent of particle size.

B. Theory for Large Particles

The theory of thermophoresis is not as well established for large aerosol particles as for small particles. The predominant theories were developed by Epstein (27), Brock (54), Jacobsen and Brock (34), Derjaguin and Bakanov (55), and Derjaguin and Yalamov (56–58). The major assumptions and problems involved in these theories relate to the description of gas molecule behavior in the neighborhood of a particle surface. The range of applicability of these theories is from the regime where $\lambda/r \ll 1$ into the intermediate regime where $\lambda/r \approx 1$. The theoretical basis for the various theories and the specific assumptions for each were recently reviewed in detail by Waldmann and Schmitt (44). A summary of the theoretical results will be presented here.

The theories of Epstein (27), Brock (54), and Jacobsen and Brock (34) are based on Maxwell's (59,60) classical boundary conditions at a particle surface and consider temperature and tangential velocity discontinuities between the gas and the surface. Epstein neglected the viscous slip portion of the tangential velocity boundary condition, while Brock included in his analysis the complete first-order Maxwellian boundary conditions. The thermal force for $\lambda/r \ll 1$, according to Epstein (27), is given by

$$\mathbf{F} = -\frac{45\pi r\eta^2\sigma}{\rho T}\left(\frac{k}{2k + k_p}\right)\nabla T \tag{5}$$

and by Brock (54) as

$$\mathbf{F} = -\frac{45\pi r\eta^2\sigma}{\rho T}\left(\frac{1}{1 + 3C_m\dfrac{\lambda}{r}}\right)\left(\frac{\dfrac{k}{k_p} + C_t\dfrac{\lambda}{r}}{1 + 2\dfrac{k}{k_p} + 2C_t\dfrac{\lambda}{r}}\right)\nabla T \tag{6}$$

where ρ is gas density, T is gas temperature, σ is the proportionality constant (Epstein and Brock use the Maxwellian approximation $\sigma = \frac{1}{5}$) for the thermal-slip portion of the tangential boundary condition (44); k_p is the thermal conductivity of the particle; and C_t and C_m are constants defined as

$$C_t = \frac{15}{8}\left(\frac{2 - \alpha_t}{\alpha_t}\right) \quad \text{and} \quad C_m = \frac{2 - \alpha_m}{\alpha_m} \tag{7,8}$$

The thermal and momentum accommodation coefficients, α_t and α_m, are defined for a point on the particle surface as

$$\alpha_t = \frac{E_i - E_r}{E_i - E_w} \quad \text{and} \quad \alpha_m = \frac{G_i - G_r}{G_i} \tag{9,10}$$

in terms of E_i, the average incident molecular energy flux; E_w, the energy flux which would be leaving the surface if the emitted molecules were in Maxwellian equilibrium with the surface; E_r, the average reflected energy flux; G_i, the average tangential component of the incident molecular momentum; and G_r, the average tangential component of the reflected molecular momentum (54).

The Epstein theory is most applicable for very small values of λ/r and for particles with low heat conductivities. Brock's equation can be seen to reduce to Epstein's equation for $\lambda/r \to 0$ with constant k/k_p. For systems with particles of higher thermal conductivities, the Epstein equation has been found to predict thermal forces many times too low.

Jacobsen and Brock (34) extended the first-order slip-flow analysis by Brock (54) and developed a second-order approximation to the thermal force. This theory is very sensitive to the choice of accommodation coefficients, and the accuracy with which they must be known is beyond current capability for their determination. Therefore, the Jacobsen and Brock equation is best used for correlation of experimental thermal-force data.

In the thermal-force theories of Epstein and Brock (eqs. 5 and 6), it is assumed that the thermal-slip proportionality constant, σ, is equal to $\frac{1}{5}$, after Maxwell (59,44). The numerical value of the constant σ depends on the distribution of gas molecule velocities near the particle surface, and Maxwell's value of $1/5$ assumes that the incident gas molecules have the same velocity distribution as in the bulk gas (57). Bakanov and Derjaguin (61) analyzed the problem of gas velocity distributions near a surface and concluded that the constant σ should have a value of about $1/175$ rather than $1/5$. This conclusion leads them to revised boundary conditions which incorporate only an isothermal velocity discontinuity at the surface.

Derjaguin and Bakanov (55) restated the conditions of incident-gas-molecule velocity distributions near a particle surface and, by using Onsager's reciprocity principle, developed a theory for the thermophoretic velocity of a particle. For the case when the temperature jump is ignored, they obtained

$$\mathbf{v} = -\frac{\eta}{\rho T}\left(\frac{1 + 8\dfrac{k}{k_p}}{1 + 2\dfrac{k}{k_p}}\right)\nabla T \tag{11}$$

Derjaguin and Yalamov (57) extended this analysis to include a temperature jump at the particle surface. They reported (58) corrections for their original analysis, and as corrected, the thermophoretic velocity is

$$\mathbf{v} = -\frac{3\eta}{\rho T}\left(\frac{\dfrac{k}{k_p} + C_t\dfrac{\lambda}{r}}{1 + 2\dfrac{k}{k_p} + 2C_t\dfrac{\lambda}{r}}\right)\nabla T \tag{12}$$

It is of interest to note that this equation is independent of the constant C_m as defined previously.

For comparison, Brock's theory may be written in terms of the thermophoretic velocity, using the drag-force equation for spherical particles

$$\mathbf{F_D} = -6\pi r \eta \mathbf{v}/C \tag{13}$$

where C is the Cunningham correction factor which interpolates between the free molecule, Epstein-drag and Stokes-drag regimes (62) and is given by

$$C = 1 + A\frac{\lambda}{r} + Q\frac{\lambda}{r}\exp\left(-br/\lambda\right) \tag{14}$$

The constants, A, Q, and b, are dependent on the gas-particle system. For oil particles in air, they have values of $1.246, 0.42$, and 0.87, respectively (63). In Brock's analysis, the thermophoretic velocity is obtained by setting the sum of the thermal and drag forces equal to zero and taking $\sigma = \frac{1}{5}$ to give

$$\mathbf{v} = -\frac{3\eta}{2\rho T}\frac{\left[1 + A\dfrac{\lambda}{r} + Q\dfrac{\lambda}{r}\exp\left(-b\dfrac{r}{\lambda}\right)\right]}{\left(1 + 3C_m\dfrac{\lambda}{r}\right)}\left(\frac{\dfrac{k}{k_p} + C_t\dfrac{\lambda}{r}}{1 + 2\dfrac{k}{k_p} + 2C_t\dfrac{\lambda}{r}}\right)\nabla T \tag{15}$$

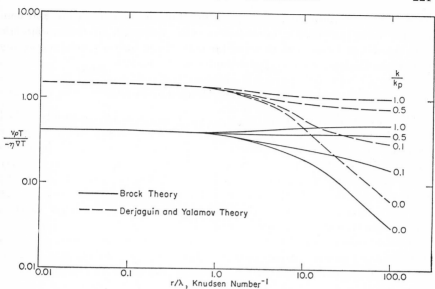

Fig. 2. Comparison of theoretical predictions of thermophoretic velocity.

Brock's theory (eq. 15) and the Derjaguin-Bakanov-Yalamov theory (eq. 12) are compared in Figure 2 in terms of a reduced thermophoretic velocity given by the ratio $-\mathbf{v}\rho T/\eta\boldsymbol{\nabla}T$. The values for the constants A, Q, and b are as given above for oil droplets in air; C_m and C_t are taken as 1.0 and 2.16 after Brock (54). The reduced thermophoretic velocity was calculated for the two theories as a function of r/λ for several values of k/k_p. It should be noted that the theoretical curves in Figure 2 are extended to lower values for r/λ than they can be expected to apply. It is evident from Figure 2 that although the dependence on r/λ is similar, the two theories are in considerable disagreement on the magnitude of the thermophoretic velocity. For small values of r/λ or small particles, comparisons can be made with the Waldmann-Bakanov and Derjaguin theory for small particles (eq. 4). For small particles, eq. 4 predicts a value for $-\mathbf{v}\rho T/\eta\boldsymbol{\nabla}T$ of $0.54 k_{tr}/k$ when a is unity. From Figure 2 it can be seen that Brock predicts a value of 0.42 and Derjaguin and Yalamov predict a value of 1.5 which indicates that the continuum theories give reasonable agreement with the free-molecule theories even though they would not be expected to apply.

It would appear that the practical solution for resolving the theoretical discrepancies is to make experimental measurements of the thermal force. As will be discussed in the following section, there are also disagreements in experimental results obtained using different experimental techniques.

There is also disagreement on what constitutes proper experimental techniques, with the result that, to date, the prediction of thermophoretic velocity and the choice of the most suitable theory for use in making such a prediction are uncertain.

C. Theory for the Transition Region

The intermediate or transition region for $\lambda/r \approx 1$ has generally defied detailed theoretical analysis. Hettner (64) developed a theory that is basically an interpolation scheme using the general theoretical behavior in the small- and large-particle regimes. The large particle analyses by Brock (54) and by Derjaguin and Yalamov (57,58) (eqs. 6 and 10) can be extended into the transition region and beyond with some success as was mentioned above. Brock (65,66), more recently, also analyzed the thermal force in the transition region ($0.2 \leq \lambda/r \leq 5$).

Brock's analysis of the transition region is, by his characterization, a nonintuitive but approximate theory. This analysis is based on a momentum-transfer calculation of the thermal force, starting with the BGK model of the Boltzmann equation for the gas-molecule velocity-distribution function near a particle surface. The form of the theoretical result is identical with an empirical representation of thermal-force data suggested by Schmitt (32). Brock's theoretical equation for the transition region is

$$\mathbf{F} = \mathbf{F}^* \exp\left(-\frac{\tau r}{\lambda}\right) \tag{16}$$

where $\mathbf{F}^*$ is the thermal force for the free-molecule or small-particle regime as given by eq. 1, and τ is a constant for each gas-particle system, independent of r/λ. For monatomic gases τ is given by

$$\tau = 0.06 + 0.09\alpha_m + 0.28\alpha_m\left(1 - \alpha_m \frac{k}{k_p}\right) \tag{17}$$

and for polyatomic gases τ is given approximately by

$$\tau = 0.06 + 0.09\alpha_m + 0.28\alpha_m\left(1 - \alpha_t \frac{k}{k_p}\right) \tag{18}$$

where α_m and α_t are the momentum and thermal accommodation coefficients defined by eqs. 9 and 10.

D. Thermal Force Experiments

Experiments for the measurement of thermal forces or thermophoretic velocities can be broadly classified into two categories and characterized by whether or not a Millikan-cell type apparatus is used. As will be shown later, specific discrepancies among experimental results can be associated with particular types of experimental systems used. However, there are no obvious experimental errors or biases inherent in any of the experimental systems. Therefore, the questions concerning the accuracy of experimental data are as yet unresolved.

The basic types of experimental apparatus used are illustrated in Figures 3a, b, and c. Figure 3a represents the Millikan-cell type apparatus with an enclosed space between two plates. The upper plate is heated and the lower plate cooled, thus providing a vertical temperature gradient and minimizing or eliminating convection currents. With this apparatus there is no gas flow while measurements are being made. An individual particle is first observed to fall under the influence of gravity in the absence of a temperature gradient; a second observation is made after a temperature gradient has been established. The particles must be charged, and they are moved vertically by an electric field imposed between the plates. The thermal force or velocity is obtained either by measuring the difference in electric-field strength required to balance the particle, or the difference in falling velocities with and without a temperature gradient. The size of each particle is determined from falling velocities with no temperature gradient. Experiments with each particle are usually run at a variety of gas pressures in order to vary the parameter λ/r. More detailed descriptions of the experimental equipment and procedures are given, for example, by Rosenblatt and LaMer (30), Schmitt (32), and Jacobsen and Brock (34).

A laminar-flow system with a vertical temperature gradient, as illustrated by Figure 3b, was used by Derjaguin and Storozhilova (67,36). A stream of particles of known size was introduced into a laminar gas stream flowing in a horizontal slit, and the thermophoretic velocity was determined from the distance the particle stream was deflected in a measured distance of travel. The velocity profiles in the laminar gas flow through the horizontal slit and the measured distances were then used to calculate the thermophoretic velocity. The upper surface was heated, the lower surface cooled, and variations in gas pressure were used to obtain various values for λ/r.

The experimental apparatus illustrated in Figure 3c was used by Derjaguin and Rabinovich (68,36) and employs a horizontal slit. In this case, there was no forced gas flow during the experimental measurements and the temperature gradient was imposed horizontally. Some horizontal gas flow of a thermo-osmotic nature exists in the system because of the temperature gradient along the horizontal slit. However, this flow was measured and accounted for in the analysis. The thermophoretic velocity was determined from measurements of the time for a particle to fall from the midplane to the lower surface of the slit. Also, individual particle velocities were measured as

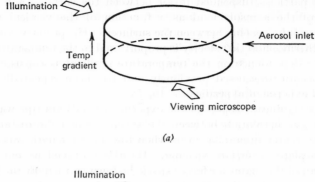

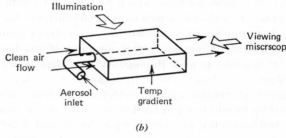

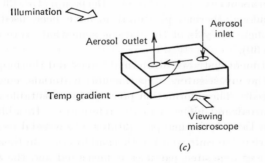

Fig. 3. Schematic illustrations of experimental methods for measuring thermal force or velocity. (a) Millikan-cell type apparatus, (b) Horizontal laminar flow apparatus with vertical temperature gradient, (c) Horizontal laminar flow apparatus with longitudinal temperature gradient.

a function of height in the slit. In some cases the particles were allowed to fall freely; in other cases a vertical electrical field was used to support the particles as they moved horizontally in the temperature gradient.

The first quantitative measurements of the thermophoretic velocity were made by Paranjpe (29) who used a static system consisting of a heated square surface (5 cm by 5 cm) held above a cooler surface. He observed a cloud of smoke particles suspended in air between the surfaces and studied the velocity of the aerosol cloud as a function of the vertical temperature gradient and the spacing between the surfaces. His primary conclusion was that the thermophoretic velocity is proportional to the temperature gradient. The linear dependence on the temperature gradient is consistent with the various theories discussed previously and has been repeatedly verified in nearly all experimental studies.

Another significant aspect of the experiments by Paranjpe was the observation of gas circulation between the surfaces when the distance between them was greater than 0.35 cm or when the surface length was less than 14 times the plate or surface spacing. He also observed no currents in the central half of the plates for larger spacings down to a length-to-spacing ratio of about 5 or 6. However, this lack of observed currents does not necessarily mean that there was no air movement in the central region. No explanation of the gas circulation was given, although it has been suggested by Derjaguin, Storozhilova, and Rabinovich (36) that it was of a thermo-osmotic nature.

Measurement of thermophoretic velocities for individual particles was first reported by Rosenblatt and LaMer (30). Their experimental apparatus was of the Millikan-cell type discussed previously and illustrated by Figure 3a. Their technique, with varying degrees of modification, has been employed in most experiments except for those by Derjaguin et al. (36,67,68). In a few cases, commercial thermal precipitators have been used in the experiments, even though analysis of the data is somewhat more difficult and less precise (37,38,39).

In Table I the experimental studies of thermal forces and thermophoresis are summarized and the investigators, experimental materials, conditions, and techniques are listed. The information presented in this table is based on measurements reported and inferred in the references. In addition to the techniques listed in Table I, thermal precipitators of several designs have been used for quantitative determinations of thermal forces. In these cases, the length of the layer of deposited particles is measured and the thermal velocity or force is calculated using an analysis of gas flow through the precipitator. Using this technique, Schadt and Cadle (37) studied tricresyl phosphate, sodium chloride, and iron particles in air; Keng and Orr (39) studied sodium chloride, aluminum oxide, magnesium oxide, iron, platinum, zinc, aluminum, and silver particles in air. Byers and Calvert (69) have also compared experimental results with theory in their studies of deposition of SiO_2, Fe_2O_3, and TiO_2 particles from turbulent air-flow through a cooled conduit.

Some of the experimental results obtained in the studies itemized in Table I are shown in Figures 4 through 7. A comparison of three different studies

TABLE I

Experimental Measurements of Thermal Force or Velocity

Reference	Particle	Gas	Particle radius range, micron	λ/r Range	k/k_p	Experimental equipment
Rosenblatt and LaMer (30)	tricresyl phosphate	air	0.4–1.6	0.04–1.5	0.12	Millikan-cell Fig. 3a
Saxton and Ranz (31)	paraffin oil	air	0.24–1.08	0.06–0.27	0.21	Millikan-cell
	castor oil	air			0.14	
Schadt and Cadle (33)	tricresyl phosphate	air	0.46–0.49	0.13–0.76	0.12	Millikan-cell
	sodium chloride	air	0.22–1.15	0.20–1.6	0.0038	
	mercury	air	0.10–1.45	0.4–3.6	0.0021	
Schmitt (32)	silicone oil, M300	argon	0.02–1.26	0.07–2.5	0.11	
	silicone oil, M300	nitrogen	0.02–1.24	0.07–2.0	0.17	
	silicone oil, M300	carbon dioxide	0.01–0.98	0.07–2.5	0.10	
	silicone oil, M300	hydrogen	0.04–0.92	0.13–1.7	1.2	Millikan-cell
	silicone oil, PH200	argon	~1.0	—	—	
	silicone oil, PH300	argon	0.715–1.24	0.1–3.2	—	
	paraffin	argon	0.7–1.2	—	—	
Jacobsen and Brock (34)	sodium chloride	argon	0.4–1.2	0.06–0.67	0.004	Millikan-cell
Gieseke (35)	tricresyl phosphate	air	0.4–0.8	0.08–1.2	0.12	Millikan-cell
Derjaguin, Storozhilova, and Rabinovich (36, 67, 68)	oil	air	0.3–0.6	0.15–3.2	0.2	horizontal slit and temperature gradient Fig. 3c
	MgO	air	0.3–0.6	0.15	—	
	tobacco smoke	air	0.3–0.6	0.15–3.4	—	
	sodium chloride	air	0.3–0.6	0.15–4.1	0.004	
	oil	helium	0.4–0.6	0.38	1.1	
	stearic acid	helium	0.3–0.6	0.38	1.1	
	oil	air	~0.25	0.27	0.2	flow cell Fig. 3b
	sodium chloride	air	~0.28	0.27–1.2	0.004	

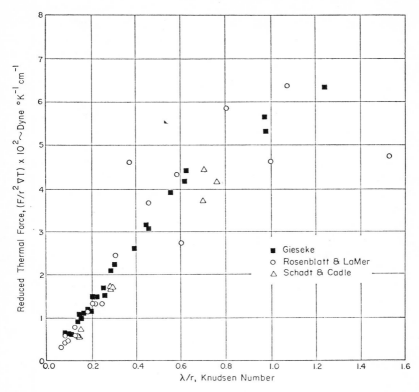

Fig. 4. Comparison of experimental measurements of the thermal force on tricresyl phosphate droplets in air. (Reproduced by permission of the Air Pollution Control Association (35).)

of tricresyl phosphate particles in air is illustrated in Figure 4. Each of these studies employed a Millikan-cell type apparatus, and the agreement among the three sets of data is quite good. It can be concluded from this comparison that the Millikan-cell technique gives reproducible and consistent measurements of the thermal force. However, several authors have suggested that the experimental results obtained with the usual Millikan-cell type apparatus may be in error (i.e., Derjaguin, Storozhilova, and Rabinovich (36)).

The major criticism of the Millikan-cell technique has been concerned with geometry, specifically the ratio of plate diameter to plate spacing. As was noted previously, Paranjpe (29) observed convection patterns between two horizontal plates when the upper plate was heated. The convection patterns disappeared when the plate diameter-to-spacing ratio has a value above about 14 and seemed to be limited to the outer regions for values of this ratio above 5 or 6. The Millikan-cell studies listed in Table I have involved plate diameter-to-spacing ratios between 6/1 and 7/1 except for a ratio of 5.1/1 as used by Rosenblatt and LaMer. In comparison, in the Derjaguin-Storozhilova-Rabinovich (36) studies, with both flowing and static gas, horizontal slits with width-to-height ratios of ≈13/1 or greater were used.

This geometry effect seems to be of less importance with the Millikan-cell than suggested, since the experiments by Rosenblatt and LaMer (30) at a ratio of 5.1/1 and those by Gieseke (35) at 7/1 are in excellent agreement. There is no question that the experimental results of Derjaguin, Yalamov, and Rabinovich with a geometry ratio of 13/1 are in some disagreement with the Millikan-cell data taken at ratios of 6/1 to 7/1, but the reasons for the differences have not yet been determined.

Comparisons of experimental results from several studies are shown in Figures 5 and 6 along with the theoretical predictions of Brock (54) and Derjaguin and Yalamov (58) for theoretical values of C_m and C_t taken as 1.0 and 2.16, respectively. The experiments for low conductivity particles, shown in Figure 5, illustrate the differences between the results obtained

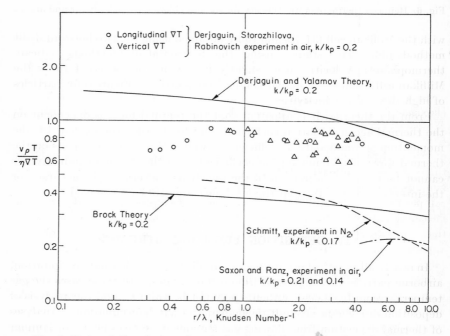

Fig. 5. Reduced thermophoretic velocity for oil droplets in air and nitrogen.

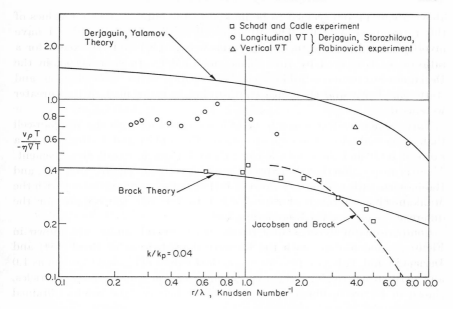

Fig. 6. Reduced thermophoretic velocity for sodium chloride particles in air and argon.

with the Millikan cell (31,32) and those obtained using the two horizontal-slit methods (36). The studies using the horizontal-slit methods indicate thermophoretic velocities from slightly less than 2 or nearly 4 times the Millikan cell results. A similar comparison is given in Figure 6 for particles of high thermal conductivities.

From the theories, it is apparent that the thermal force is dependent on the thermal accommodation coefficient, and for Brock's theory (eq. 13) the momentum accommodation coefficient as well. However, the effects on the thermal force of accommodation coefficients and the thermal conductivities cannot be resolved entirely until the questions concerning the accuracy or the interpretation of the experimental measurements are resolved.

IV. DEPOSITION FROM GAS STREAMS

In nearly all practical occurrences or applications of thermal precipitation, airborne particles are deposited from a flowing gas. In these cases the gas temperature and velocity gradients may not be uniform, and the analysis of deposition rate will be strongly dependent on their determination. Analyses of thermal deposition from flowing gas streams have been made for laminar and turbulent flow but for only a few simple geometries.

A. Deposition from Laminar Flow

For all treatments of thermal deposition from laminar flow, the temperature gradient, and hence the thermophoretic velocity have been assumed
constant.

For fully developed laminar flow between parallel plates with constant
cross-section, and with no temperature gradient the velocity distribution can
be written as

$$U(y) = U_{max}\left[1 - \frac{(y-h)^2}{h^2}\right] \tag{19}$$

where h is half the distance between the plates and $y = 0$ corresponds to the
lower plate as shown in Figure 7. For a constant temperature gradient
between the plates, the thermophoretic velocity, $\mathbf{v}$, is given by $(y_2 - y_1)/\theta$,
where $(y_2 - y_1)$ represents the vertical distance travelled in a time interval θ.

The horizontal distance a particle moves along with the gas in the time
interval θ is

$$x = \frac{U_{max}}{\mathbf{v}}\int_{y_1}^{y_2}[1 - (y-h)^2/h^2]\,dy \tag{20}$$

which when integrated gives

$$x = \frac{U_{max}}{\mathbf{v}}\left[\frac{(y_2{}^2 - y_1{}^2)}{h} - \frac{(y_2{}^3 - y_1{}^3)}{3h^2}\right] \tag{21}$$

The case of most practical interest is when the entire cross-sectional area at
the entrance is filled with an aerosol and the length required for complete
deposition is sought. This length corresponds to conditions when $y_2 = 2h$
and $y_1 = 0$, and is given by

$$x = \frac{4U_{max}h}{3\mathbf{v}} \tag{22}$$

The thermophoretic velocity, $\mathbf{v}$, can be specified by one of the equations
previously given (i.e., eq. 12 or eq. 15) if a precipitator is to be designed.

In some instances, analyses of the type described above have been used to
determine the thermophoretic velocity. This has been done by Derjaguin

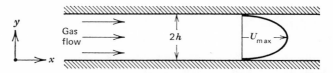

Fig. 7. Coordinate system for parallel plates, fully developed laminar flow.

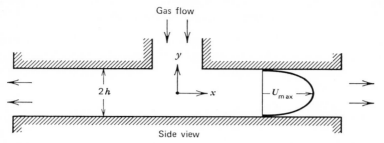

Fig. 8. Coordinate system for parallel plates, gas flow radially outward.

and Storozhilova (67,36). In the case of a flowing gas stream, the aerosol deflection below the center line, L, and the distance in the direction of flow, x, required for this deflection were used to give the thermophoretic velocity with the equation

$$\mathbf{v} = \frac{LU_{\max}}{x}\left(1 - \frac{L^2}{3h^2}\right) \tag{23}$$

Westerboer (70) has used a similar analysis for deposition in a thermal precipitator having a heated ribbon placed in the center of the horizontal slit. The effect of a temperature gradient on the velocity profile was apparently assumed to be negligible in each case.

One type of thermal precipitator is constructed of parallel circular plates and gas enters at the center of the plates and flows radially outward. A diagram of the vertical cross-section of this type of precipitator is given in Figure 8. Considering both thermal and gravitational forces on the aerosol particles as they move horizontally, an equation for the width of the particle deposition pattern can be developed. Such an analysis was reported by Gordon (71) using Epstein's (27) equation for the thermal force. The corrected deposition equation* is given as

$$\frac{q}{\pi z(z + d)} = \frac{2r^2(\rho_p - \rho)g}{9\eta} + \frac{3}{4}\left(\frac{R\eta}{M\rho}\right)\left(\frac{k}{2k + k_p}\right)\frac{\Delta T}{h} \tag{24}$$

where q = volumetric gas flow rate; d = diameter of inlet; z = length of deposit; g = acceleration of gravity; ρ_p = density of particle; R = gas constant; M = molecular weight of gas; ΔT = temperature difference between plates; and $2h$ = spacing between plates. Binek (73) and Keng and Orr (39) have used this equation in different forms and have also included the effect of velocity slip for small particles.

---

* In the equation given by Gordon (71), the particle diameter incorrectly appeared to the first power rather than squared. In the quotation of this equation by Green and Lane (72), this original error is repeated.

B. Deposition from Turbulent Flow

Thermal precipitation of aerosols in sampling lines, on heat-transfer surfaces, and from most natural-convection flows of hot gases is most likely to occur from a turbulent gas because these flows are most often turbulent. In these cases, the controlling step will usually be the thermophoretic movement of the particles through the laminar sublayer near the surface since the eddy diffusivity will be large enough to transport particles to this sublayer at a comparatively high rate. Experiments by Byers and Calvert (69) have shown that for their conditions of high temperature gradients neither turbulent diffusivity in the bulk gas nor Brownian diffusion in the laminar sublayer is a significant factor in the deposition rate. For large particles, inertial deposition could become significant (74).

An analytical model was developed by Postma (19) for predicting particle deposition from turbulent gas flows in a circular conduit. A driving force for thermal precipitation exists when the gas stream is at a higher temperature than the conduit wall. In this model a fully developed turbulent flow is assumed, such that there is a turbulent core region from the center of the conduit to near the wall ($y^+ \leqslant 5$) and a laminar sublayer from there to the surface. In the turbulent core region the eddy diffusivity is assumed to be great enough so that thermal deposition through the laminar sublayer is controlling and the concentration profile is flat up to the laminar sublayer. In the sublayer the Reynolds' analogy for heat- and momentum-transfer is used to predict the temperature gradient from the velocity profile. For the case considered, a constant wall temperature was assumed and the thermophoretic velocity in the radial direction determines the particle deposition rate.

With the above assumptions, Postma has obtained the temperature gradient across the laminar sublayer starting with a heat balance on an element as shown in Figure 9. In general, Reynolds' analogy and Prandtl's extension of Reynolds' analogy (75) were followed. Since the Prandtl number for most gases is near unity, the velocity and temperature gradients are related by (19)

$$\frac{U_\delta}{\overline{U}} = \frac{T_\delta - T_w}{T - T_w} \tag{25}$$

and from the heat balance

$$T_\delta - T_w = \frac{U_\delta}{\overline{U}}(T_0 - T_w) \exp\left(-\frac{2fx}{d_w}\right) \tag{26}$$

where T_δ and U_δ are the temperature and velocity at the outer edge of the laminar sublayer; T_w and T_o are the wall and entering gas temperatures;

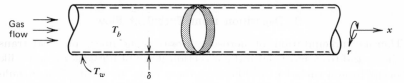

Fig. 9. Coordinate system for turbulent flow through a circular conduit.

$\overline{U}$ is the average gas velocity through the conduit of diameter d_w, x is the axial distance; and f is the friction factor. The temperature gradient in the laminar sublayer is given as

$$-\frac{dT}{dr} = \frac{N_R f (T_0 - T_w)}{2 d_w} \exp\left(-\frac{2fx}{d_w}\right) \tag{27}$$

where N_R is the Reynolds number for flow through the conduit.

A mass balance on the particulate matter gives a concentration gradient,

$$\frac{dN}{dx} = \frac{4yN}{\overline{U} d_w} \tag{28}$$

where N is the particle concentration. The theoretical expressions for the thermophoretic velocity of particles can be generalized as

$$\mathbf{v} = -K_T \frac{\eta}{\rho T} \frac{dT}{dr} \tag{29}$$

Inserting eqs. 27 and 29 into eq. 28 gives

$$\frac{dN}{dx} = \frac{2 K_T f (T_o - T_w)}{T_\delta d_w} \exp\left(-\frac{2fx}{d_w}\right) \tag{30}$$

Using the axial temperature dependence of the sublayer boundary at $y^+ = 5$ and the Prandtl relationship giving the gas velocity at this same radial location as $U_\delta/\overline{U} = 5\sqrt{f/2}$, the temperature T_δ obtained using eq. 26 becomes

$$T_\delta = T_w + 5 \sqrt{\frac{f}{2}} (T_o - T_w) \exp\left(-\frac{2fx}{d_w}\right) \tag{31}$$

Equation 30 can now be integrated between $N = N_o$ and $N = N$ for $x = 0$ to $x = x$ giving

$$\ln\left(\frac{N}{N_o}\right) = -\frac{6 K_T f}{5 d_w \sqrt{f/2}} \left\{ x - \frac{d_w}{2f} \ln \frac{\left[5\sqrt{\frac{f}{2}}(T_o - T_w) + T_w \exp\left(\frac{2fx}{d_w}\right)\right]}{\left[5\sqrt{\frac{f}{2}}(T_o - T_w) + T_w\right]} \right\} \tag{32}$$

Byers and Calvert (68) have essentially followed Postma's analysis for particle deposition from turbulent flow but have included several modifications. In their analysis, the Reynolds' analogy or Stanton number definition, $H/\overline{U}\rho C_p d_w = f/2 =$ Stanton number, was not employed. Therefore, the arguments of the exponential functions in eqs. 26, 27, and 30 were retained as $-4Hx/\overline{U}\rho C_p d_w$, where H is the heat-transfer coefficient at the tube wall. Further, Brock's equation for the thermal force was taken in the form

$$\mathbf{F} = \frac{-12\pi\eta r^2 \left(C_{tm}\dfrac{\lambda}{r}\right)\left(\dfrac{k}{k_p} + C_t\dfrac{\lambda}{r}\right)}{\left(1 + 3C_m\dfrac{\lambda}{r}\right)\left(1 + 2\dfrac{k}{k_p} + 2C_t\dfrac{\lambda}{r}\right)} \nabla T \tag{33}$$

where C_{tm} is a thermal creep first-order coefficient. Brock (76) has shown that C_{tm} is probably not much different from the value given by

$$C_{tm} = \frac{3}{4}\sqrt{\frac{2R}{\pi T}} = \frac{3\eta}{4\rho\lambda T} \tag{34}$$

where R is the gas constant. When this definition is used, it can be shown that eqs. 33 and 6 are equivalent.

Byers and Calvert (69) have used eq. 33, thus assuming no axial temperature dependence for C_{tm}. Neglect of the axial changes in C_{tm} is probably not serious considering the accuracy of the theories and the fact that other gas and particle physical properties are assumed constant, including the other coefficients C_t and C_m. Byers and Calvert also assume that for their experimental conditions and conduit length, the abrupt change in wall temperature at the beginning of their test section results in the average radial temperature gradient being 2 times that obtained from the heat balance. For these assumptions they obtain a particle concentration dependence as

$$\ln\left(\frac{N}{N_o}\right) = \frac{-2\rho C_p f N_R K^1}{d_w H}(T_o - T_w)\left[1 - \exp\left(-\frac{4Hx}{\overline{U}\rho C_p d_w}\right)\right] \tag{35}$$

where K^1 is defined by

$$K^1 = \frac{\left(1 + A\dfrac{\lambda}{r}\right)(C_{tm}\lambda)\left(\dfrac{k}{k_p} + C_t\dfrac{\lambda}{r}\right)}{\left(1 + 3C_m\dfrac{\lambda}{r}\right)\left(1 + 2\dfrac{k}{k_p} + 2C_t\dfrac{\lambda}{r}\right)}$$

and $1 + A(\lambda/r)$ is the linear part of the velocity slip correction as given in eq. 14.

Byers and Calvert (69) reported experimental measurements of particle deposition from a hot gas onto the water-cooled walls of a circular tube. They used sodium chloride particles with diameters between 0.3 and 1.3 microns and reported data for Reynolds' numbers ranging from 8750 to 24,040, Knudsen's numbers (λ/r) from 0.1 to 0.35, and inlet gas temperatures from 450° to 900° F. The collection efficiency by thermal deposition was found to increase with increasing Knudsen's number, inlet temperature, and tube length. The effect of Reynolds' number was comparatively small except for the larger particles. Most significant was the increase in collection efficiency with decreasing particle size. The analysis given by eq. 35 shows some efficiency increase with decreasing particle size, but the measured effect is much greater than predicted. When the many assumptions and approximations included in the theoretical analysis are considered, the agreement between theory and experiment is surprisingly good, especially for the larger particle sizes studied.

V. THERMAL PRECIPITATION FOR AEROSOL SAMPLING

The thermal deposition of aerosol particles from a flowing gas stream has been exploited most successfully in thermal precipitators and aerosol sampling. The first practical use of thermal deposition as a sampling technique was apparently made by Lomax, who, as reported by Whytlaw-Gray and Patterson (77), used this technique to concentrate dust particles on a glass slide. In this application the glass slide was placed on a cooled lower plate and the dusty gas was passed between it and a heated upper plate. The most widely used precipitator design can be attributed to Green and Watson (43,24) and consists of a heated wire suspended between cool glass slides.

In most applications of thermal precipitators, the intended methods of analysis for collected particles include chemical and microscopic examination. The considerable interest in thermal precipitators for sampling has apparently resulted from the development of the electron microscope which allows sizing of the very small particles which are collected by thermal precipitation. The ability of a thermal precipitator to collect very small particles has usually been considered a significant advantage, although the presence of small particles normally contained in the ambient atmosphere can complicate the analysis for particles from a specific source. Most thermal precipitators operate at efficiencies approaching 100% for particles with diameters below about 10 microns, and the collection should be essentially complete for particles down to molecular sizes.

Several precipitator designs have evolved which have various advantages. These designs, their specific characteristics, and some general problems associated with thermal precipitators will be discussed below. However, all well-designed thermal precipitators can be characterized by their high efficiencies for small particles, low volumetric sampling rates, and compatibility with microscopic particle-sizing techniques.

A. Thermal Precipitator Designs

The first successful thermal precipitator evolved from the work of Whytlaw-Gray, Lomax, Green, and Watson (78) and can be called a wire-and-plate design. This precipitator incorporates an experimental technique of Aitken (2) and was proposed as a practical device by Green and Watson (43,24). The basic design, illustrated in Figure 10, consists of a narrow flow channel bounded by glass slides on either side of an electrically heated wire held perpendicular to the direction of gas flow. The glass slides are held in contact with metal heat sinks and are so positioned that they intercept the dust-free space surrounding the hot wire. Temperature gradients, usually about 4000 to 6000° C/cm, are established between the wire and the glass deposition surfaces. Sampling rates usually range from 2 to 10 cm³/min. Hasenclever (79) found collection efficiencies for this precipitator to be essentially 100% for particles smaller than 5 microns and Watson (80) found efficiencies in the range from 90% to 100% for particles smaller than 10 microns.

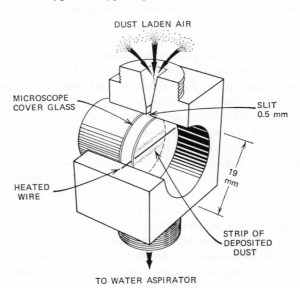

Fig. 10. Standard wire-and-plate design for a thermal precipitator. (British Crown Copyright. Reproduced by permission of Her Majesty's Stationery Office.)

Alterations of the wire-and-plate precipitator design by various experimentors have usually been directed toward alleviating specific sampling and analysis problems characteristic of this design. Two general problems associated with particles larger than 10 microns are reduced collection efficiency and clogging of the narrow flow passages. Drinker and Hatch (81) suggest shielding the precipitator inlet from settling particles, and they report that Green used a wire mesh across the precipitator inlet. Both procedures would keep larger particles from entering the thermal precipitator. Burdekin and Dawes (82) devised a simple impaction or settling stage for the precipitator inlet which collects large particles. Similarly, Hamilton (83) used an inertial plus gravitational settling scheme to remove larger particles ahead of the thermal precipitation section. In other precipitator designs, such as Wright's (84), the dusty gas flows horizontally between flat plates, and hence, there is both gravitational settling and thermal deposition onto the lower plate. In designs of this type (84–87), collection of larger particles occurs simultaneously with the collection of smaller particles.

Various design improvements have been suggested to allow for higher flow rates (88,89), more efficient use of the electrical energy (89), better centering of the heating wire in the flow channel (88,90), or to permit longer sampling periods (83,85,91). Cember, Hatch, and Watson (91) devised a precipitator with a rotating collection plate. Although this permits longer sampling times by allowing replacement of the glass collection slides at less frequent intervals, it has been suggested (92) that gas leakage is a significant problem with this device. Orr and Martin (85) designed a continuous precipitator with deposition on a moving plastic tape. In this precipitator the dusty air flows inward from the periphery of the space between two circular plates. The upper plate is heated and the gas flow exits at the center of the upper plate. The plastic deposition tape is held in contact with the bottom, cooler plate and is advanced through the precipitator mechanically. The plates are concave downward so that the plastic tape can be held snuggly against the lower plate.

Another advancement in thermal precipitator design has been made at the Institute for Measurement and Control of the Technical University in Berlin (93). The application in this case is directed at gravimetric measurements of the particle content of a gas rather than the usual application of thermal precipitators as sample-collectors for microscopic examination. In this device the particles are deposited on a thin quartz plate by thermal precipitation and the collected particle mass measured with an electromagnetic ultra-microbalance. The sequence of collection, quartz-plate cleaning, and microbalance rezeroing is automatically performed by program control.

A significant analysis problem is associated with the wire-and-plate design

in that a gradation of particle sizes occurs along the collection plate in the
direction of gas flow. The small particles are concentrated nearer the
entrance, with a shift to larger particles farther downstream as was noted by
Fuchs and Ianovskii (94), Ashford (95), Thürmer (96), and Polydorova (97).
Two basic solutions have been employed. First, the heating wire or the
deposition slides have been made to oscillate back and forth; this causes a
central portion of the deposit to be formed from all particle sizes. Second, a
wider deposition area such as under a heating plate has been used.

Oscillating wires or plates have been used, for example, by Walton, Faust,
and Harris (98), Kitto (99), Walton (100), and Poppoff (101). Oscillation is
usually over only a few millimeters. Under certain conditions, this oscillatory
motion can affect deposition efficiency as noted by Poppoff (101) who found
that when the velocity of oscillation exceeded 1 cm/sec, collection efficiency
was reduced. Poppoff (101) reported that the oscillating precipitator which
he used was designed by Wilson (102) as an improvement of the design by
Walton, Faust, and Harris (98) and that it gave collection efficiencies of
100 % for the size range 1 to 4 microns, 99 % for the range 4 to 6 microns, and
98 % for the range 6 to 10 microns.

Particle deposition over a broader area along the direction of the gas
flow has been achieved by several methods. Brunetti, Magill, and Sawyer
(103) suggested that a heating ribbon be used to replace the usual heating
wire. Walkenhorst (104) designed a precipitator using such a heating

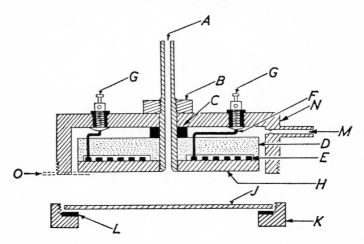

Fig. 11. Gravimetric thermal precipitator. A, inlet tube; B, locking nut; C, plastic
sleeve; D, asbestos disc; E, heater element; F, connection to heater; G, terminals; H,
aluminum hot plate; J, collecting plate; K, locking ring; L, rubber washer; M, outlet
tube; N, aluminum case; O, 0.015 in. gap between hot and cold plates. (Reproduced
by permission of *Science* (84).)

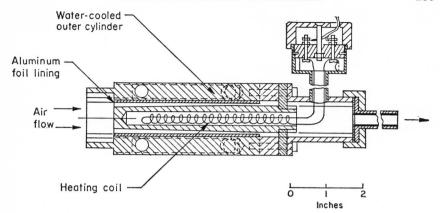

Fig. 12. Annular thermal precipitator (107). (Reproduced by permission of the authors, The Institute of Physics, and the Physical Society.)

ribbon and reported on its performance. This concept has also been used by Schadt and Cadle (37) and analyzed theoretically by Westerboer (70). In an analogous design, Thürmer (89) used several coiled heated wires arranged side by side. The logical extension of this general concept leads to a precipitator so constructed that the gas passes between heated and cooled plates. A design of this type was originally used for sampling by Lomax (78) and was adopted by Harrington and Crozier (105), Kethley, Gordon, and Orr (86), Wright (84), Homma, Koshi, and Sakabe (87), and Binek (73). An example of this type of precipitator, the design by Wright (84), is shown in Figure 11. A modification of the two-plate design has been reported by Guintini and Godard (106) who used a heated, hollow needle through which the dusty gas passed prior to impingement on a cooled collection surface.

The feasibility of particle deposition within concentric cylinders was demonstrated by Aitken (2) and Bancroft (40) and employed by Bredl and Grieve (107) in the design of the precipitator as shown in Figure 12. This precipitator maintains the gas flow through an annulus around a central, heated cylinder within a cooled outer cylinder.

In the precipitator designs with large, heated surfaces, the temperature gradient is often much smaller than when a heated wire is used. However, the path length for deposition in the direction of gas flow is much greater. In some cases greater gas-flow rates are possible and rates up to 1 liter/min have been used successfully as compared with the usual flow rate of a few milliliters/min for the wire-and-plate design.

The use of a broader deposition area has helped overcome another major disadvantage of the thermal precipitator. When the original wire-and-plate design is used to deposit particles on an electron microscope grid, local variations in the temperature gradient cause particles to be deposited preferentially over the metal grid structure. This nonuniform deposition pattern was first noted by Cartwright (108) and confirmed, using radioactive aerosols, by Billings, Megaw, and Wiffen (109). The two-plate precipitators give a more uniform deposit on microscope grids (87). However, for either the wire-and-plate or two-plate designs, a more suitable procedure is to allow the particles to deposit on a uniform substrate or film of carbon or Formvar. Such a technique has been suggested and found successful by Cartwright (108) and by Fuchs and Iankovskii (110). After the particles have been deposited, the substrate is removed with the particle deposit intact, and transferred to a microscope grid for examination.

B. Thermal Precipitator Characteristics and Uses

The characteristics of thermal precipitators employed as sampling devices have been discussed in the preceding section to the extent that they have affected precipitator designs and design modifications. However, several additional characteristics and their affect on the applicability of thermal precipitators to various sampling situations are of importance.

As was mentioned previously, the major use of samples collected in thermal precipitators has been for microscopic examination. In most cases, particle size and concentration in the sampled gas are determined. Any particle counting procedure involves problems, many of which are related to human errors (95). Beadle and Kerrich (111) have made a statistical analysis of the sources and magnitudes of errors in thermal precipitator samples. Roach (112) has analyzed the variability of particle counts obtained from his samples and estimates the total standard error in any count to be about 15%. Another common problem is the expected shift in measured particle size as the sampling velocity becomes anisokinetic. However, Hodkinson, Critchlow, and Stanley (113) found no significant variation in sampling efficiency for two standard precipitator designs when the sampling was performed in air streams flowing at velocities between 40 and 1200 ft/min.

One source of errors which is characteristic of all collected particle samples is the overlapping of particles on the collection surface. If many particles are collected on a small deposition area, it is impossible to determine which particles were agglomerates in the air before deposition and which particles became agglomerated on the surface as they were deposited. This general problem and methods for estimating the associated errors have been analyzed theoretically by Armitage (114) and Irwin, Armitage, and Davies (115). The overlap error has been studied experimentally by Roach (116) and Hodkinson (117) who found good agreement with theory. For the standard wire-and-plate thermal precipitator, the overlap error can be kept below 5%

by following guidelines established by Davies (118) and adhering to his recommended sample volumes.

With thermal precipitators the collected particles are randomly oriented with respect to the deposition surface. This is true whether the deposit consists of single particles or agglomerates. This characteristic of particle orientation has been reported by Donoghue (119), Watson (120), and Cruise (121). Donoghue and Cruise used shadowing techniques to enable them to view the particle dimensions perpendicular to the collection surface. Watson made comparisons between particles collected with a conifuge and with a thermal precipitator. Particles deposited by inertial forces, either by gravity, in a conifuge, or with an impinger, tend to orient themselves so they are stable aerodynamically. Usually they are deposited with a larger-than-average projected area perpendicular to the collection surface. If total particle mass is to be determined from the particle deposit, random orientation is preferable. However, if aerodynamic or Stokes' size is of interest, random orientation is less desirable.

Another characteristic of importance in particle deposition is the magnitude of the thermophoretic velocity. This, of course, reflects the ease or rapidity of particle deposition. As was discussed previously, the major factors, once the temperature gradient is fixed, are the thermal conductivity of the particle and its surface properties. As illustrated in Figure 2, the thermophoretic velocity can be expected to increase as the thermal conductivity of the particle decreases. Keng and Orr (39), however, found this effect to be small and report that particle deposition in a thermal precipitator is nearly independent of the thermal conductivity of the particles. For sampling in normal atmospheres, surface properties of the particle can be expected to be relatively unimportant since most particles will have various gases and vapors adsorbed on their surfaces and, consequently, will tend to behave similarly, regardless of particle material. Particles of widely differing materials have been collected successfully in thermal precipitators including metals (39), metal oxides (38,39,73,87,94,97), quartz (94,122), sodium chloride (37,39,73,87,94), and dust from pulverized coal (108,122). Airborne dust in coal mines (99,123,124), atmospheric particles (123,125,126), smoke (107,123), organic droplets (37), and microorganisms (127) have also been collected.

Extensive comparisons of particle concentrations and sizes obtained from samples collected by thermal precipitation and by various other methods have been made by Hasenclever (79), Avy, Benarie, and Hartogensis (128), and Davies, Alyward, and Leacey (122). The major conclusion of Davies, Alyward, and Leacey is that in comparison to thermal precipitators, impingement devices tend to give erroneously high particle counts for smaller particles when agglomerates are sampled. They attribute this to breakup of the agglomerated particles during the impingement process. Hence, an important characteristic of thermal precipitation is that sampled particles are deposited so gently that they do not break apart. Therefore, the deposited particles, as subsequently observed, are in the same form as they were while airborne.

The advantages and disadvantages of thermal precipitators can be summarized as:

1. they have low gas-flow sampling rates, usually in the range from 5 to 10 cm^3/min but for some designs up to 1 liter/min;

2. collection efficiency is high for particles from 10 microns in diameter down to molecular sizes;

3. collection efficiencies are lower for particles larger than 10 microns;

4. deposition is gentle and agglomerates are not broken;

5. orientation of the deposited particles is random;

6. some gradation of particles by size occurs which must be accounted for in a size analysis and/or in the choice of precipitator design.

In general, thermal precipitators are well suited for collection of particles to be observed with an optical or electron microscope.

VI. THERMAL PRECIPITATION FOR AIR CLEANING

The elimination of air pollution, preservation of clean and healthy environments, and protection of contamination-sensitive processes are accomplished by a variety of air-cleaning methods. These air-cleaning procedures are most effectively applied to dusty effluent gases prior to their release into the atmosphere or to dusty inlet air before use. The air-cleaning devices usually used on either effluent or inlet gases have been most often classified as mechanical collectors, wet scrubbers, filters, or electrostatic precipitators. There has been, to the knowledge of this author, no practical application of thermal precipitation as an air-cleaning technique, although several air-cleaner designs have been proposed.

In any particle-collection device, the removal of particles from a gas is dependent on several mechanisms for moving the particles through the gas to the deposition surface. The deposition surface may be in one of a variety of forms, such as solid walls, water drops, or fibers, depending on the collector type. In the four general collector classifications listed above, the deposition mechanisms result from Brownian diffusion and from electrostatic and inertial forces. Particle collection is also affected by interception in which case the particle follows a streamline in the gas and deposition results when a particle passes the collection surface at a distance less than the particle radius.

It is most likely that thermal precipitation is not well suited for large-scale air-cleaning applications except for special situations. For example, when the gas to be cleaned is at an elevated temperature and the gas can be cooled, thermal deposition can occur without a large expenditure of energy other than the sensible heat of the gas. In other cases when proper conditions exist, combinations of thermal deposition and other basic deposition mechanisms can lead to substantial improvements in collector performance.

A. Thermal Collectors

Several air-cleaner designs have been proposed which are based entirely on particle deposition by thermal forces. The first such suggestion was made by Aitken (2) who proposed using the heat contained in effluent combustion gases to collect particles from these gases. Aitken experimented with a "thermic filter" which consisted of concentric cylindrical tubes with the inner tube cooled and the outer tube heated; dusty gases were made to flow in the annulus between the tubes. His success in collecting particles with the device led him to design and successfully test a metal chimney to be placed over a smoky paraffin lamp. This chimney design is shown in Figure 13a. The inner surface of the annulus was heated by the effluent gas; the outer boundary of the annular space, where deposition occurred, was cooled by exposure to room air.

Aitken suggested that the collection principles demonstrated in his experiments could be successfully applied for the collection of dust or soot from the combustion products in large chimneys. He proposed a multistage thermal precipitator where cycling of the effluent gas flow could be used to reheat the stack gases after they cooled in the thermal precipitators. Reheating would allow the buoyancy of the exhaust gases to be partially restored.

In his search for methods of collecting or eliminating chemical warfare smokes, Bancroft (40) investigated the possibility of using thermal precipitators. His experimental equipment followed Aitken's concentric-cylinder design but had the inner cylinder heated and the outer cylinder cooled. He found that to maintain collection efficiencies near 100%, the flow rate could be increased with increasing temperature gradient. The relationship between flow rate and temperature was found to be nearly linear. With tubes 30 in. long, collection was essentially complete for a flow rate of 550 cm³/min and a temperature difference between inner and outer cylinders of 80° C.

Blacktin (41) experimented with several precipitator designs using a substantially different technique. The basis of Blacktin's designs is a heated wire mesh through which the air is drawn. The heated mesh sets up a

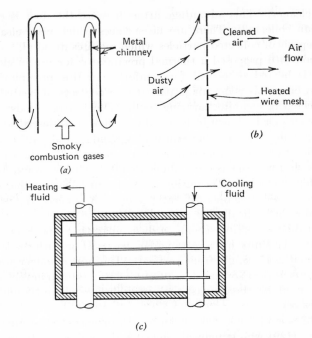

Fig. 13. Air cleaner designs.   (a) Aitken's chimney for a paraffin lamp, (b) Blacktin's heated-mesh concept, (c) Sherwood's cleaner for circulating room air.

temperature gradient so that dust particles approaching the mesh are repulsed.   As a result, the air passing through the wire mesh is cleaned.   The dust apparently remains with a convective flow of waste gas over the heated mesh surface.   The operation of these air cleaners is illustrated in Figure 13b. Blacktin employed conduction, radiation, and electrical energy for heating the wire mesh.   With the wire mesh radiantly heated to about 85° C, a particle removal efficiency of 94% was measured for an air flow of 4 ft³/hour through an effective mesh area of 0.2 sq in. with a pressure loss of 1 mm of water.

The high collection efficiencies reported by Blacktin are rather surprising. If the temperature gradient is assumed to be effective over one-half the aperture width in 60 I.M.M. mesh, the thermophoretic velocity can be estimated as about 1 cm/sec or less.   A thermophoretic velocity of 1 cm/sec should be near the maximum attainable for the experimental conditions, but it is still much less than the superficial face velocity of the gas flow into the mesh of about 24 cm/sec.   Hence, the apparent effectiveness of this air-cleaner design cannot be explained on the basis of mesh temperature alone.

Blacktin's concept has one very important advantage when applied as an inlet air cleaner. This advantage arises because the dust is concentrated rather than collected. There are no surfaces which need cleaning, nor is there a deposit of collected particles which requires disposal.

Sherwood (40) proposed a thermal precipitator for use in cleaning room air as it is heated or cooled for comfort. In this precipitator, thermal deposition occurs as room air is passed between sets of heated and cooled fins connected to conduits carrying cooling and heating fluids. The design of this air cleaner is shown in Figure 13c. Sherwood described the case for air heating in most detail, where the cooling fluid is intake air for a combustor and the heating fluid is the combustion exhaust. For the case where air cooling is desired, the cooling fluid would be below room temperature. Particle deposition would occur on the cooler surfaces in either case. If an air cleaner of this type operates satisfactorily, it is necessary to remove the collected dust periodically.

Another dust collection scheme, a dust-separating pipe, was suggested by Clusius (9). This pipe has a rectangular cross-section with the longer dimension at about a 45° angle from vertical. The upper, longer wall is heated and the lower cooled so that a convective flow is set up in the cross-section. This flow circulates the air past the cooler surface where thermal deposition of particles occurs.

The reduction of fumes from open-hearth furnaces was studied by Strauss and Thring (129) who employed packed beds as filters. In most cases the effect of temperature difference between the gas and the packing was obscured by the fact that the packing materials were heated by the gas in a relatively short time compared with the collection times. However, Thring and Strauss (130), in reviewing the effects of high temperatures on particle collection, report that experiments have shown that when a packed bed is initially cold the collection is more nearly complete. Strauss (131) estimated that if a temperature difference of 50° C can be maintained between the dusty gas and a 9-in.-deep packing, a collection efficiency of 98.8 % can be achieved for 0.1-micron particles. The suggested method for cooling the packed bed is to use it as an air preheater in which another gas stream is heated as it cools the bed (130). A cycling system of this type was also proposed by Aitken (2). When a packed bed is incorporated to give higher collection efficiencies, the proposed scheme shows some promise.

The qualities of any air-cleaning device which allow it to be successfully used include initial cost, operating and maintenance costs, collection efficiency, and compatability with the gas to be cleaned. Because there are so many diverse applications for air cleaners, and because the initial cost of a thermal precipitator is expected to be comparable with that of other collection equipment, the critical factors for a general assessment of thermal-precipitator applicability are the operating costs and collection efficiency.

Operating costs are generally dependent on utility and maintenance requirements. For a thermal precipitator at this stage of development, it is desirable to compare collection efficiency with the power required to achieve that efficiency. Some elementary comparisons of this type are made below.

To operate a thermal precipitator, it is necessary to provide a source of heat; this is the major energy requirement. For the simple case of laminar gas flow between heated and cooled plates, the heat required can be estimated as that transferred across the gas stream between the plates. If the energy required for gas flow, and if radiation heat-transfer and sensible heat change of the gas are neglected, then a simple conduction problem can be formulated. On this basis and using a value for $-\mathbf{v}\rho T/\eta\nabla T$ of 0.2, it can be determined that a thermal precipitator should require about 0.3 watt for a gas flow of 100 ml/min. This can be compared with the 10 watts for 100 ml/min reported by Wright (84) for his parallel-plate design and the 1 watt for 6–7 ml/min he reports for the Green and Watson (42,43) design. In these latter cases, some sensible heat change of the gas occurs and this, in part, accounts for the higher energy requirements.

A comparison is also possible with other dust collectors. Semrau (132) has collected data on the efficiencies and power requirements for wet scrubbers and from these data it can be estimated that for a collection efficiency of 99.3%, the power required is about 10 hp per 1000 ft³ gas/min. The comparable figures for the ideal thermal precipitator are an efficiency of 100 percent for particles smaller than 5 microns with a power requirement of about 100 hp per 1000 ft³ gas/min. On the basis of this comparison, it can be concluded that the power required for thermal precipitation would be at least an order of magnitude greater than that required for most other air cleaners. However, unlike the case with most other collectors, the power required for collection decreases as particle size decreases.

A more economical adaptation of thermal precipitation for air cleaning is possible where hot gases are to be cleaned. In such a case the sensible heat of the gas provides the energy source, and if the gas is cooled with proper heat-transfer surfaces, deposition will occur. This basic technique is essentially that proposed by Aitken (2) and Thring and Strauss (130), and studied by Byers and Calvert (69) for thermal deposition from turbulent flows.

The case of turbulent flow through a pipe can be analyzed using eq. 35. If it is assumed that the pipe is long enough so that the inlet gas temperature is reduced to the wall temperature, the maximum efficiency from thermal deposition can be calculated. With an inlet gas temperature of 900° F, a wall temperature of 80° F, and a thermophoretic velocity given by $-\mathbf{v}\rho T/\eta\nabla T = 0.2$, the maximum collection efficiency is about 54%. Some improvement in this efficiency can be realized for particles smaller than about 1 micron. The natural logarithm of the maximum efficiency, for this case, is proportional to the difference between the inlet and wall temperatures. For the temperature conditions considered, the power required, in the form of heat lost from the gas, is about 230 hp/1000 cfm. This power requirement would be excessive if it must be supplied for particle collection only; however, if it is waste or usable heat energy, the major considerations would be pressure loss through the pipe.

The only economical applications of thermal precipitation for air cleaning seem to be where there is heat energy available in the gas to be cleaned. It is possible that this heat energy can be used in some air-preheater arrangement. If such an arrangement is to be devised, it would be advantageous to use a better contactor than an open pipe, perhaps a packed bed as suggested by Thring and Strauss (130).

B. Thermal Deposition as a Contributing Mechanism

In many dust collectors, more than one mechanism is responsible for collection of dust particles. For example, filtration involves inertial deposition, interception, and Brownian diffusion. The mechanisms operate most effectively on different particle sizes, but there is some overlap. Likewise, other air cleaners are dependent on particle deposition by more than one basic mechanism. It is in this area of multiple deposition mechanisms that thermal forces may contribute to dust collection in a variety of types of collection equipment.

Wet scrubbers often give higher collection efficiencies when the scrubbing liquid is cold, as noted by Collins (133), Hansen (134), Lapple and Kamack (135), Semrau, Marynowski, Lunde, and Lapple (136), and Litvinov (137). The reason for the improved performance may be partially the result of reduced pressure loss in some cases (138), but some efficiency increase can also be expected (137). Two mechanisms which can be postulated as causes for the increased particle deposition rates are thermophoresis and diffusiophoresis with a condensing vapor flow. Semrau (139) and Lapple and Kamack (135) have discussed deposition mechanisms for such conditions and Litvinov (137) has analyzed the case for water vapor condensation. When particle deposition is caused by vapor condensing on a cool surface, Goldsmith and May (140) report that the combination of deposition mechanisms can be quite effective. The combined thermal and diffusion-induced particle movement exceeds that expected from the sum of forces for the two mechanisms.

The deposition of particles from a hot gas in a cooled, packed bed is again a combination of mechanisms. The usual mechanisms causing deposition in filtration processes will occur in any case, but the addition of thermal forces may substantially improve collection. Although there is very little available information on thermal deposition combined with other mechanisms, the few but favorable results seem to warrant more extensive investigations.

VII. SUMMARY

Aerosol particles suspended in a gas in which there exists a temperature gradient, experience a thermal force directed toward the cooler temperatures. This force is the result of energy transferred from the suspending gas to the particles. Thermophoretic movement of particles in the direction of the heat flux near a cool surface can result in deposition of particles on the surface.

Theoretical analyses and experimental measurements of thermal forces have been made for a range of particle sizes and numerous particle materials. Theoretical and experimental results are in good agreement for small particles, but differences among theories and among experimental results are prevalent for larger particles.

Thermal deposition frequently occurs on various cooling surfaces and in most instances is a nuisance. However, thermal precipitators have successfully exploited the process of thermal deposition as a sampling technique. Several air-cleaner designs based on thermophoretic effect have been suggested; however, in practice thermal deposition has not been used for this purpose. Nevertheless, some promise is seen for thermal air cleaners when deposition is from a hot gas or when thermal forces can contribute as a supplementary deposition mechanism.

VIII. ACKNOWLEDGMENTS

The author wishes to thank Dr. A. K. Postma and Dr. W. G. N. Slinn of The Pacific Northwest Laboratories of Battelle for reviewing this report and for their many helpful comments and suggestions. Financial support for manuscript typing and figure preparation provided by the Columbus Laboratories of Battelle is gratefully acknowledged.

References

1. Brock, J. R., *J. Colloid and Interface Sci.*, **25**, 564 (1967).
2. Aitken, J., *Trans. Royal Soc. Edinburgh*, **32**, 239 (1884).
3. Lodge, O. J., *Nature*, **29**, 610 (1884).
4. Gibbs, Wm. E., *Clouds and Smokes*, Blakistons and Sons, Philadelphia (1924).
5. Poynting, J. H., and Thomson, J. J., *Textbook of Physics, Vol. IV, Heat*, C. Griffin and Co., Ltd., London (1906).
6. Hooper, W. J., *Physics*, **1**, 61 (1931).
7. Bonnell, D. G. R., and Burridge, L. W., England Building Research Bulletin, No. 10 (February, 1931) [Department of Scientific and Industrial Research].

8. Nielsen, R. A., *Heating, Piping, and Air Conditioning*, **12,** 389 (1940).

9. Clusius, K., *Z. VDI—Beiheft Verfahrenstechnik*, **2,** 23 (1941).

10. Bryhni, A., *Electrotecknisk Tidsskrift* (Norway), **74,** No. 7, 103 (1961).

11. Roots, W. K., and Walker, F., *IEE Proceedings* (Great Britain), **112,** No. 3, 511 (1965).

12. Drake, D. F., and Harnett, C. G., *J. Inst. Fuel*, **42,** 339 (1969).

13. Hawes, R. I., and Garton, D. A., *Chem. Proc. Engr.*, **48** (8), 143 (1967).

14. Boothroyd, R. G., *Chem. Proc. Engr.*, **50** (10), 108 (1969).

15. Schluderberg, D. C., Whitelaw, R. L., and Carlson, R. W., *Nucleonics*, **19,** No. 8, 67 (August 1961).

16. Abel, W. T., O'Leary, J. P., Bluman, D. E., and McGee, J. P., *Experiments with Solids-in-Gas Suspensions as Heat Transport Mediums, U.S. Bureau of Mines, Report of Investigations, 6255* (1963).

17. Wachtel, G. P., and Waggener, J. P., *Flow Stability of Gas-Solids Suspensions, NYO-2974-1* (February, 1964).

18. Postma, A. K., and Schwendiman, L. C., *Studies in Micromeritics I. Particle Deposition in Conduits as a Source of Error in Aerosol Sampling, HW-65308* (1960).

19. Postma, A. K., *Studies in Micromeritics II. The Deposition of Particles in Circular Conduits Due to Thermal Gradients, HW-70791* (1961).

20. Tyndall, J., *Proc. Royal Inst.* (London), **6,** 3 (1870).

21. Rayleigh, Lord, *Proc. Royal Soc.*, **34,** 414 (1882).

22. Ibid, *Nature*, **28,** 139 (1882).

23. Lodge, O. J., and Clark, J. W., *Phil. Mag.*, **17,** 214 (1884).

24. Watson, H. H., *Trans. Faraday Soc.*, **32,** 1073 (1936).

25. Miyake, S., *Report of the Aeronautical Research Institute, Tokyo University*, **10,** 85 (1935).

26. Zernik, W., *Brit. J. Appl. Phys.*, **8,** 117 (1957).

27. Epstein, P. S., *Z. Phys.*, **54,** 537 (1929).

28. Cawood, W., *Trans. Faraday Soc.*, **32,** 1069 (1936).

29. Paranjpe, M. K., *Proc. Indian Acad. Sci.*, **4a,** 423 (1936).

30. Rosenblatt, P., and LaMer, V. K., *Physical Review*, **70,** 385 (1946).

31. Saxton, R. L., and Ranz, W. E., *J. Appl. Phys.*, **23,** 917 (1952).

32. Schmitt, K. H., *Z. Naturforsch.*, **14a,** 870 (1959).

33. Schadt, C. F., and Cadle, R. D., *J. Phys. Chem.*, **65,** 1689 (1961).

34. Jacobsen, S., and Brock, J. R., *J. Colloid and Interface Sci.*, **20,** 544 (1965).

35. Gieseke, J. A., *J. Air Poll. Control Assoc.*, **18,** 682 (1968).

36. Derjaguin, B. V., Storozhilova, A. I., and Rabinovich, Ya. I., *J. Colloid and Interface Sci.*, **21,** 35 (1966).

37. Schadt, C. F., and Cadle, R. D., *J. Colloid Sci.*, **12,** 356 (1957).

38. Orr, C., Jr., and Wilson, T. W., *J. Colloid Sci.*, **19,** 571 (1964).

39. Keng, E. Y. H., and Orr, C., Jr., *J. Colloid and Interface Sci.*, **22,** 107 (1966).

40. Bancroft, W. D., *J. Phys. Chem.*, **24,** 421 (1920).

41. Blacktin, S. C., *J. Soc. Chem. Ind.*, **58,** 334 (1940); **59,** 153 (1940).

42. Sherwood, T. K., *United States Patent 2,833,370* (May 16, 1958).

43. Green, H. L., and Watson, H. H., *Medical Research Council, Special Report No. 199*, H.M.S.O. (1935).

44. Waldmann, L., and Schmitt, K. H., in *Aerosol Science*, C. N. Davies, editor, Academic Press, New York (1966).

45. Einstein, A., *Z. Physik*, **27,** 1 (1924).

46. Clusius, K., *Z. VDI—Beiheft Verfahrenstecknik*, **2,** 23 (1941).

47. Stetter, G., *Staub*, **20**, 244 (1960); and *Anz. Oesterr. Akad. Wiss. Math. Naturw. Kla.*, **7** (1951) as cited by Thurmer, H., *Staub*, **20**, 6 (1960).

48. Derjaguin, B. V., and Bakanov, S. P., *Kolloid Zh.*, **21**, 377 (1959).

49. Waldmann, L., *Z. Naturforsch.*, **14a**, 589 (1959).

50. Mason, E. A., and Chapman, S., *J. Chem. Phys.*, **36**, 627 (1962).

51. Monchick, L., Yun, K. S., and Mason, E. A., *J. Chem. Phys.*, **39**, 654 (1963).

52. Epstein, P. S., *Phys. Rev.*, **23**, 710 (1924).

53. Slinn, W. G. N., Shen, S. F., and Mazo, R. M., *J. Stat. Phys.*, **2**, 251 (1970).

54. Brock, J. R., *J. Colloid Sci.*, **17**, 768 (1962).

55. Derjaguin, B. V., and Bakanov, S. P., *Dokl. Akad. Nauk SSSR.*, **147**, 139 (1962).

56. Derjaguin, B. V., and Yalamov, Yu. I., *Dokl. Akad. Nauk SSSR.*, **155**, 886 (1964).

57. Derjaguin, B. V., and Yalamov, Yu. I., *J. Colloid Sci.*, **20**, 555 (1965).

58. Derjaguin, B. V., and Yalamov, Yu. I., *J. Colloid and Interface Sci.*, **22**, 195 (1966).

59. Maxwell, J. C., *Philosophical Transactions of the Royal Society*, **170**, No. 7 of Part 1, 231 (1879).

60. Kennard, E. H., *Kinetic Theory of Gases*, McGraw-Hill Book Company, Inc., New York (1938).

61. Bakanov, S. P., and Derjaguin, B. V., *Dokl. Akad. Nauk SSSR*, **139**, 71 (1961).

62. Cunningham, E., *Proc. Royal Soc.*, **83A**, 357 (1910).

63. Fuchs, N. A., *The Mechanics of Aerosols*, The Macmillan Co., New York (1964).

64. Hettner, G., *Z. Physik*, **37**, 179 (1926).

65. Brock, J. R., *J. Colloid and Interface Sci.*, **23**, 448 (1967).

66. Brock, J. R., *J. Colloid and Interface Sci.*, **25**, 392 (1967).

67. Derjaguin, B. V., and Storozhilova, A. I., *Kolloidnyi Zhurnal*, **26**, No. 5, 583 (1964).

68. Derjaguin, B. V., and Rabinovich, Ya. I., *Kolloidnyi Zhurnal*, **26**, No. 5, 649 (1964).

69. Byers, R. L., and Calvert, S., *Ind. Eng. Chem. Fundamentals*, **8**, 646 (1969).

70. Westerboer, I., *Staub*, **21**, 466 (1961).

71. Gordon, M. T., *Science*, **119**, 816 (1954).

72. Green, H. L., and Lane, W. R., *Particulate Clouds: Dusts, Smokes, and Mists*, Second Edition, E. & F. N. Spon Ltd., London (1964).

73. Binek, B., *Staub*, **25**, 261 (1965).

74. Friedlander, S. K., and Johnstone, H. F., *Ind. Eng. Chem.*, **49**, 1151 (1957).

75. Knudsen, J. G., and Katz, D. L., *Fluid Dynamics and Heat Transfer*, McGraw-Hill Book Co., Inc., New York (1958).

76. Brock, J. R., *J. Phys. Chem.*, **66**, 1763 (1962).

77. Whytlaw-Gray, R., and Patterson, H. S., *Smoke*, Edward Arnold and Co., London (1932).

78. Whytlaw-Gray, R., Lomax, R., Green, H. L., and Watson, H. H., *British Patent* 445551 (1936).

79. Hasenclever, D., *Staub*, **41**, 388 (1955).

80. Watson, H. H., *Brit. J. Appl. Phys.*, **9**, 78 (1958).

81. Drinker, P., and Hatch, T., *Industrial Dust*, Second Edition, McGraw-Hill Book Co., Inc., New York (1954).

82. Burdekin, J. T., and Dawes, J. G., *Brit. J. Industr. Med.*, **13**, 196 (1956).

83. Hamilton, R. J., *J. Sci. Instruments*, **33**, 395 (1956).

84. Wright, B. W., *Science*, **118**, 195 (1953).

85. Orr, C., and Martin, R. A., *Review Sci. Instr.*, **29**, 129 (1958).

86. Kethley, T. W., Gordon, M. T., and Orr, C., Jr., *Science*, **116**, 368 (1952).

87. Homma, K., Koshi, S., and Sakabe, H., *Bull. Nat. Inst. Industrial Health* (Japan) **7,** 25 (1962).
88. Hasenclever, D., *Staub*, **22,** 99 (1962).
89. Thürmer, H., p. 237 in *Aerosols, Physical Chemistry and Applications*, K. Spurny, editor, Gordon and Breach, Science Publishers, Inc., New York (1965).
90. Donoghue, J. K., *J. Sci. Instruments*, **30,** 59 (1953).
91. Cember, H., Hatch, T., and Watson, J. A., *AIHA Quarterly*, **14,** 191 (1953).
92. Billings, C. E., and Silverman, L., *J. Air Poll. Control Assoc.*, **12,** 586 (1962).
93. Anonymous, *Staub-Reinhalt, Luft* (in English), **27,** No. 9, 36 (1967).
94. Fuchs, N. A., and Iankovskii, S. S., *Dokl. Akad. Nauk SSSR*, **119,** 1177 (1958).
95. Ashford, J. R., *Brit. J. Appl. Phys.*, **11,** 13 (1960).
96. Thürmer, H., *Staub*, **20,** 6 (1960).
97. Polydorova, M., *Staub-Reinhalt Luft*, **27,** 345 (1967).
98. Walton, W. H., Faust, R. C., and Harris, W. J., *A Modified Thermal Precipitator for the Quantitative Sampling of Aerosols for Electron Microscopy, Porton Tech. Paper No. 1* (March 31, 1947).
99. Kitto, P. H., *Bulletin of the Institution of Mining and Metallurgy, No. 497,* 1 (1948).
100. Walton, W. H., p. 45 in article edited by D. G. Drummond, *J. Royal Micros. Soc.*, **70,** 1 (1950).
101. Poppoff, I. G., *AIHA Quarterly*, **15,** 145 (1954).
102. Lauterbach, K. E., Wilson, R. H., Laskin, S., and Meier, D. W., *Design of an Oscillating Precipitator, Univ. of Rochester Atomic Energy Project Report UR-199.*
103. Brunetti, C., Magill, P. L., and Sawyer, F. G., p. 643 in *Air Pollution, Proceedings of the United States Tech. Conf. on Air Poll.*, L. C. McCabe, Chairman, McGraw-Hill Book Co., Inc., New York (1952).
104. Walkenhorst, W., *Staub*, **22,** 103 (1962).
105. Harrington, E. R., and Crozier, W. D., *A Thermal Precipitator for Sparse Dispersions of Aerosols, Report 2-NR,* The New Mexico School of Mines (December 1948).
106. Guintini, J., and Godard, L., *Geofisica Pura E Appl.*, **50,** 42 (September–December 1961).
107. Bredl, J., and Grieve, T. W., *J. Sci. Instruments*, **28,** 21 (1951).
108. Cartwright, J., *Brit. J. Appl. Phys., Supp. No. 3,* S109 (1954).
109. Billings, C. E., Megaw, W. J., and Wiffen, R. D., *Nature*, **189,** 336 (1961).
110. Fuchs, N. A., and Iankovskii, S. S., *Kolloidnyi Zhurnal*, **21,** 133 (1959).
111. Beadle, D. G., and Kerrich, J. E., *J. Chemical Metall. and Mining Soc. of S. Africa* **56,** 219 (December 1955).
112. Roach, S. A., *Brit. J. Industrial Med.*, **16,** 104 (1959).
113. Hodkinson, J. R., Critchlow, A., and Stanley, N., *J. Sci. Instruments*, **37,** 182 (1960).
114. Armitage, P., *Biometrika*, **36,** 257 (1959).
115. Irwin, J. O., Armitage, P., and Davies, C. N., *Nature*, **163,** 809 (1949).
116. Roach, S. A., *Brit. J. Industrial Med.*, **15,** 250 (1958).
117. Hodkinson, J. R., *Annals Occup. Hyg.*, **6,** 131 (1963).
118. Davies, C. N., *Dust is Dangerous*, Faber and Faber Limited, London (1954).
119. Donoghue, J. K., *Brit. J. Appl. Phys., Supp. No. 3,* S94 (1954).
120. Watson, H. H., *Brit. J. Appl. Phys., Supp. No. 3,* S94 (1954).
121. Cruise, A. J., *Engineering*, **183,** 366 (1957).
122. Davies, C. N., Alyward, M., and Leacey, D., *Archives Industr. Hyg. and Occup. Med.*, **4,** 354 (1951).

123. Cartwright, J., Nagelschmidt, G., and Skidmore, J. W., *Royal Meteorology Society Quarterly J.*, **82**, 82 (1956).
124. Cartwright, J., and Skidmore, J. W., *Fuel*, **36**, 205 (1957).
125. Froula, H., Bush, A. F., and Bowler, E. S. C., *Proceedings of the Third National Air Pollution Symposium*, p. 102 (1955).
126. Bush, A. F., *A.M.A. Arch. Ind. Health*, **15**, 1 (1957).
127. Orr, C., Jr., Gordon, M. T., and Kordecki, M. C., *Appl. Microbiol.*, **4**, 116 (1956).
128. Avy, A. P., Benarie, M., and Hartogensis, F., *Staub-Reinhalt. Luft* (English Translation), **27**, No. 11, 1 (1967).
129. Strauss, W., and Thring, M. W., *J. Iron and Steel Inst.*, **196**, 62 (1960).
130. Thring, M. W., and Strauss, W., *Trans. Inst. Chem. Engrs.*, **41**, 248 (1963).
131. Strauss, W., *Industrial Gas Cleaning*, Pergamon Press Ltd., Oxford (1966).
132. Semrau, K. T., *J. Air Poll. Control Assoc.*, **10**, 200 (1960).
133. Collins, T. T., Jr., *Paper Industry and the Paper World*, **29**, Nos. 5, 6, and 7, 830 and 984 (1947).
134. Hansen, G. A., *J. Air Poll. Control Assoc.*, **12**, 409 (1960).
135. Lapple, C. E., and Kamack, H. T., *Chem. Engr. Progress*, **51**, No. 3, 113 (1955).
136. Semrau, K. T., Marynowski, C. W., Lunde, K. E., and Lapple, C. E., *Ind. Engr. Chem.*, **50**, 1615 (1958).
137. Litvinov, A. T., *Zh. Prikl. Khim.*, **40**, 353 (1967).
138. Gieseke, J. A., *Pressure Loss and Dust Collection in Venturi Scrubbers*, Ph.D. Thesis, Univ. of Washington (1963).
139. Semrau, K. T., *J. Air Poll. Control Assoc.*, **13**, 587 (1963).
140. Goldsmith, P., and May, F. G., in *Aerosol Science*, C. N. Davies, editor, Academic Press, New York (1966).

Nomenclature

A — Constant in equation for frictional-slip correction

a — Fraction of molecules diffusely emitted from particle surface

b — Constant in equation for frictional-slip correction

C — Correction factor for frictional slip

C_m — Coefficient of thermal slip

C_p — Heat capacity

C_t — Coefficient of momentum slip

C_{tm} — First-order coefficient of thermal creep

$\bar{c}$ — $\sqrt{8KT/\pi m}$ = Mean thermal velocity of the gas molecules

d — Inlet tube diameter

d_w — Conduit diameter

E_i — Average incident molecular energy flux at a point on the particle surface

E_r Average reflected energy flux from a point on the particle surface

E_w Energy flux which would be leaving the particle surface if the emitted molecules were in Maxwellian equilibrium with the surface

f Friction factor

$\mathbf{F}$ Thermal force

$\mathbf{F}^*$ Thermal force for $\lambda/r \to \infty$

$\mathbf{F_D}$ Drag force on a moving particle

G_i Average tangential component of the incident molecular momentum at a point on the particle surface

G_r Average tangential component of the reflected molecular momentum at a point on the particle surface

g Acceleration of gravity

H Heat transfer coefficient

h One-half the distance between two parallel plates

K Boltzmann constant

K^1 Constant defined for Equation 36

K_T Constant in generalized equation for the thermophoretic velocity

k Thermal conductivity of gas

k_p Thermal conductivity of particle

k_{tr} $15K\eta/4m =$ Translational contribution to thermal conductivity of gas

L Distance from centerline between parallel plates

M Molecular weight

m Molecular mass

N Particle concentration

N_o Initial particle concentration

N_R Reynolds number

p Gas pressure

Q Constant in equation for frictional-slip correction

q	Volumetric gas flow rate
R	Gas constant
r	Particle radius
T	Absolute gas temperature
ΔT	Temperature difference between parallel plates
T_o	Initial gas velocity
T_w	Wall temperature
T_δ	Temperature at outer edge of the laminar sublayer
U	Gas velocity
$\bar{U}$	Average gas velocity
$U_{\max}$	Maximum gas velocity in a parabolic velocity distribution
U_δ	Velocity at outer edge of the laminar sublayer
$\mathbf{v}$	Thermophoretic velocity of particle
x	Distance in direction of gas flow
y	Distance perpendicular to direction of gas flow
z	Length of particle deposit
α_m	Momentum accommodation coefficient
α_t	Thermal accommodation coefficient
η	Gas viscosity
θ	Time interval
λ	Mean free path of gas molecules
ρ	Gas density
ρ_p	Particle density
σ	Thermal-slip proportionality constant
τ	Constant in thermal-force equation for the transition region.

The Literature of Air Pollution

HAROLD M. ENGLUND

Air Pollution Control Association
Pittsburgh, Pennsylvania

I. INTRODUCTION: THE LITERATURE OF AIR POLLUTION

The literature dealing with air pollution began as a trickle and is presently a wide, rushing river which occasionally threatens to become a flood.

This body of literature began in the seventeenth century when John Evelyn presented his thoughtful treatise on London air to Charles II.

Today, published material on air pollution is both technical and nontechnical; it comprises books, manuals, conference proceedings, bibliographies, abstract publications, professional society journals, commercial magazines, and, finally, the Monday morning newsletters.

In this final chapter of the second volume of a series on air pollution control, why do we concern ourselves with the literature of air pollution?

First, even though this series contains a wealth of information both through the data presented and in the bibliographic references, additional reading will be warranted in selected areas for many individuals.

Second, every printed word represents the past through a reporting of research completed, work done, and thoughts crystallized. Even as one reads these words, further research is taking place, work is being carried out, and thoughts are jelling to become part of tomorrow's literature on air pollution. We must, therefore, identify the media which provide a current awareness of developments in this fast-moving field.

The following pages contain categorized listings of publications, as well as technical information resources, which it is hoped will satisfy both the retrospective and the current needs in air pollution literature.

II. TECHNICAL INFORMATION RESOURCES

Air Pollution Control Association (APCA)
4400 Fifth Avenue, Pittsburgh, Pennsylvania 15213

The Air Pollution Control Association is a 64 year old technical association whose activities are directed to the collection and dissemination of authoritative information about air pollution and its control. Media utilized by the Association include its monthly Journal (described in Section IV), special publications, and national and sectional meetings.

The Association also publishes manuals containing informative reports prepared by its technical committees in their various fields.

Other materials available from the Association include back issues of the APCA Journal, preprints of papers delivered at its annual meetings, and reprints of papers published in its journal. Finally, the Association produces annually a directory of governmental air pollution control agencies.

Bay Area Air Pollution Control District (BAAPCD)
999 Ellis Street, San Francisco, California 94109

The Technical Library of the Bay Area Air Pollution Control District, located in San Francisco, contains approximately 14,000 items, almost all on 35 mm microfilm. New references are added at the rate of 600–700 per year. It is indexed by keywords, using a Uniterm card system for searching by visual comparison to identify common document numbers. Since the process becomes cumbersome for a collection of this size the index is currently being converted to a Termatrex system.

Since this is one of the first major air pollution document collections available, numerous subscribers have purchased sets of the films and index cards and receive semi-annual updates. These subscribers include state and local control agencies, universities, and federal government agencies. The collection may be consulted at other locations as well as at the Bay Area District.

Center for Air Environment Studies
The Pennsylvania State, University
University Park, Pennsylvania 16802

As an adjunct to its educational, research, and training programs, the Center for Air Environment Studies operates a substantial air pollution information service. Its collection, now totaling over 7500 documents, primarily serves the faculty and students on campus, but access by others is available on request. A Termatrex system is used for searching. The Center also subscribes to the Bay Area library collection, which contains over 14,000 documents, as well as up to 40 specialized journals.

Air Pollution Titles is published every 2 months as a current-awareness tool, cumulating the year's entries in the November–December issue. In format, it is a computer-produced Key-Word-in-Context (KWIC) index similar to *Chemical Titles*, and covers over 1000 journals in the air pollution and related fields with a maximum lapse of time of 8–10 weeks.

An *Index to Air Pollution Research* was published in July of 1966, 1967, and 1968 also in the KWIC format. The index covers nonprofit research in the field and the results of a survey of air pollution research projects being conducted by industrial, sustaining, and corporate members of the Air Pollution Control Association and the American Industrial Hygiene Association.

Clearinghouse for Federal Scientific and
Technical Information
U.S. Department of Commerce
Springfield, Virginia 22151

Although the Clearinghouse, a unit of the Department of Commerce, is not a specialized information service it is the sales outlet for copies of government-sponsored technical reports available to the general public. Thus it is the source for copies of such reports identified and announced by the various specialized services as pertinent to air pollution. These are available both as hard bound copies and as microfiche.

The Clearinghouse also publishes *Clearinghouse Announcements in Science and Technology (CAST)* in 46 separate subject areas. The semi-monthly issues are in a format permitting rapid scanning of report titles and also provide abstracts and other information. Among the subject areas carrying material related to air pollution are: No. 3 Area Development Planning; No. 5 Atmospheric Sciences; No. 9 Chemical Processing; No. 37 Power Source Devices; and No. 38 Propulsion Systems.

Graphic Arts Technical Foundation
4615 Forbes Avenue
Pittsburgh, Pennsylvania 15213

The Graphic Arts Technical Foundation, located in Pittsburgh, has established a Pollution Information Center to collect technical and administrative information on pollution problems, primarily air, of interest to the printing and metal decorating industries. Initial impetus came from the National Metal Decorators Association, whose members are concerned with compliance with the increasing number of codes concerning solvent emissions.

The Center gathers information on pollution problems and control methods peculiar to the industry and provides a technical advisory service, including literature searches, on a fee basis. Air pollution abstracts are published as a section in the Foundation's monthly *Graphic Arts Abstracts*. The publication is available to nonmembers of the Foundation by subscription, and photocopies of abstracted papers may be purchased.

Institute of Paper Chemistry
Lawrence University
Appleton, Wisconsin 54911

The information service of The Institute for Paper Chemistry is an example of an industry-oriented service which includes in its broader scope a thorough coverage of air pollution aspects.

Its monthly *Abstract Bulletin* (*ABIPC*), now in its 40th volume, covers all technical areas of interest to the pulp and paper industry. Air Pollution problems are largely limited to kraft pulping operations. However, references of air pollution interest are spread among several of the subject categories into which the bulletin is organized. In order to insure complete coverage, it is necessary to refer to several headings in the subject index included in each issue and to the accompanying keyword supplement.

The Abstract Bulletin is available on a subscription basis. Photocopies of abstracted articles and translations are sold, as well as specialized bibliographies.

Office of Technical Information and Publications
Environmental Protection Agency (EPA)
Air Pollution Control Office
Research Triangle Park
North Carolina 27709

The Office of Technical Information and Publications (OTIP) provides a complete system of technical information and communication activities for the Federal EPA Air Pollution Control Office (APCO). The principal element of APCO theretofore charged with the responsibility of technical communication has been the Air Pollution Technical Information Center (APTIC). The establishment of OTIP suggests an expansion of The Clean Air Act of 1963 which delegated to the Secretary of the Department of Health, Education, and Welfare the authority to "collect and disseminate, in cooperation with other public or welfare agencies, institutions, and organizations having related responsibilities basic data on chemical, physical, and biological effects of varying air quality and other information pertaining to air pollution and the prevention and control thereof."

The overall objective of OTIP is to provide timely technical information and publication services to each program functioning in APCO and to maintain intermediary technical communication among all agencies, government or private, involved in the quest for clean air. Specifically OTIP is charged with the following:

1. An effective information and retrieval system responsive to the needs of APCO technical personnel.

2. Production, acquisition, and dissemination of all forms of recorded media related to air pollution technical information including books, periodicals, pamphlets, microforms, photographs, film strips, motion pictures, video tapes, and other audiovisual materials.

3. Provision of publication and editorial services for all APCO programs.

4. Provision of complete graphic art services for APCO.

5. Arrangement and management of technical conferences and seminars on air pollution.

OTIP has the responsibility of screening two sources of technical literature, i.e., journal literature and report literature. The journal literature screening involves 1004 journals, 392 of which are published in English. The dilution from an original screening list of 4300 foreign and domestic journals is due to an evaluation of those journals that yielded the most fruitful technical articles dealing with the air pollution aspects of medicine, chemistry, meteorology and engineering. The report literature phase of the screening program is conducted by OTIP technical information specialists trained in the technical disciplines of biology, chemistry, meteorology and engineering. Bulletins from Government clearinghouses are the primary sources of the report literature.

The gathered information is microformed for storage and dissemination and simultaneously formatted for manipulation on a central computer system which is under the control of the Environmental Health Service (EHS). Requests for information are answered using the computer data bank of current air pollution technical literature. This bank of literature is updated by approximately 800 articles per month so the most pertinent current literature is always available. A monthly abstract bulletin including bibliographic information and delineating the availability of the document is distributed to technically-oriented organizations and individuals.

III. BOOKS, MANUALS, PROCEEDINGS, AND BIBLIOGRAPHIES

A. General Aspects

Air Conservation: The Report of the Air Conservation Commission of AAAS published by the American Association for the Advancement of Science, 1515 Massachusetts Avenue, N.W., Washington, D.C. 20005; 348 pages; 1966.

Prepared for broad though not necessarily mass distribution, this final report of the AAAS Air Conservation Commission is designed to be useful to scientists in a wide variety of disciplines and to laymen interested in the subject of air conservation.

Air Pollution(2nd Edition); edited by Arthur C. Stern; published by Academic Press, 111 Fifth Avenue, New York, N.Y. 10003; 3 volumes; 1968.

This is the most comprehensive work in the field. Volume I deals with air pollution and its effects, including meteorology; Vol. II with analysis, monitoring and surveying; and Vol. III with sources of air pollution and their control, including a section on legislation, standards, information and education.

Air Pollution and Its Control, by Wayne T. Sproull; published by Exposition Press Inc., 50 Jericho Turnpike, Jericho, N.Y. 11753; 106 pages, 1970.

This examination of air pollution, written in layman's language by a nationally-known authority, clearly answers the questions concerned people have about this urgent problem. What is air pollution? How does it endanger health? What is it costing to control it? How effective are present control methods? What does the future hold? City planners, public officials, citizen's committees, and others actively working to solve the air pollution problem will find many answers in this book.

Air Pollution Control; by W. L. Faith, published by John Wiley & Sons, Inc., New York, N.Y.; 260 pages; 1959.

This book is directed to those who want a better understanding of the more common air pollution problems, the available means for assessing the magnitude of the problem, and appropriate methods of abatement.

Air Pollution Control, Part I; edited by W. Strauss; published by Wiley-Interscience, a Division of John Wiley & Sons, Inc., New York, N.Y.; 1971.

This book contains a series of seven chapters on critical problem areas in air pollution. They include dispersion of pollutants, formation and control of oxides of nitrogen, control of sulphur oxides and motor car exhausts, and the mechanisms of collection in electrostatic precipitators, fiber filters, and scrubbers.

Air Pollution Control, Part II, edited by W. Strauss; published by Wiley-Interscience, a Division of John Wiley & Sons, Inc., New York, N.Y.; 1971.

The second book of the series contains chapters on the abundance and fate of pollutants, the philosophy of pollution control legislation, control of pollution from nuclear reactors, thermal precipitation, particle clouds, costing of gas cleaning plant, and this review of the literature in the field.

Air Pollution Handbook, edited by P. L. Magill, F. R. Holden and C. Ackley; McGraw-Hill, New York, N.Y.; 1956.

This handbook gives a fairly comprehensive coverage of similar topics to the work edited by Dr. Stern. It is now, however, about 13 years old, and so tends to be somewhat out of date on many aspects.

Air Pollution Experiments for Junior and Senior High School Science Classes; published by the Air Pollution Control Association, Pittsburgh, Pa.; 64 pages; 1969.

This manual contains 19 air pollution experiments which are designed to portray air pollution phenomena, including effects, measurement, analysis, and meteorological aspects.

Air Pollution Publications 1966–1968 (PHS Publication No. 979), is available from the Superintendent of Documents, U.S. Government Printing Office, Washington, D.C. 20402; 522 pages; 1969.

A bibliography with abstracts of air pollution publications by National Air Pollution Control Administration personnel and personnel of organizations receiving Federal air pollution funds covers such aspects of the problem as the sources of air pollution and their control, effects of pollutants on man, fauna, flora, and materials, air quality standards, legal and social aspects, and basic science and technology.

The Air We Live In by James Marshall; published by Coward-McCann, Inc., 200 Madison Avenue, New York, N.Y. 10016; 96 pages; 1969.

This book is the first in "The New Conservation Series" initiated for "age twelve and over" readers. The series deals directly with man and his relationship to his environment. The book deals with the reasons for and means of controlling the air we breathe.

Atmospheric Pollution (*Its Origin and Prevention*), by A. R. Meetham; a Pergamon Press book published in the U.S.A. by The Macmillan Company, 60 Fifth Avenue, New York, N.Y.; 302 pages; 1964.

This third edition of a book first published in 1952 is a readable, practical presentation of pollution problems and their control, in terms of sources, distribution, measurement, effects and legislation.

The Breath of Life, by Donald E. Carr; published by W. W. Norton and Company, Inc., New York, N.Y.; 175 pages; 1965.

Air pollution has been with us since the discovery of fire, but the major culprit now, according to Donald E. Carr, a professional research chemist from Los Angeles, is the burning of gasoline and high-sulfur fuels.

Cleaning our Environment: The Chemical Basis for Action; published by the American Chemical Society, Washington, D.C.; 250 pages; 1969.

This report is an objective account of the current status of the science and technology of environmental improvement. It concludes with a series of recommendations of measures which, if adopted, should help to accelerate the sound development and use of that science and technology. Principal focus is on chemistry and chemical engineering.

Clean the Air; by Alfred E. Lewis; published by McGraw-Hill Book Company, 330 W. 42 St., New York, N.Y.; 96 pages; 1965.

Admittedly addressed to young people, 12 years old and upward, an informative book by Alfred Lewis, an editor of National Petroleum News, also serves all ages as a short, convenient guide to the current status of air pollution and its control in this nation.

Crisis In Our Cities; by Lewis Herber; published by Prentice-Hall, Inc., Englewood Cliffs, N.J.; 239 pages; 1965.

The tremendous increase in urban population, according to Lewis Herber, clearly justifies the warning that next to the question of keeping world peace, metropolitan planning is the most serious single problem faced by man in the second half of the twentieth century. His new book takes a hard look at the effects of the modern metropolis on human physical and mental health.

Congress and the Environment; edited by Richard A. Cooley and Geoffrey Wandesforde-Smith; published by University of Washington Press, Seattle, Washington 98105; 277 pages, 1970.

The volume consists of a series of original case studies which developed out of a year-long environmental policy seminar held at the University of Washington.   Each chapter surveys a recent piece of legislation to determine how Congress has handled a particular environmental problem.   Focusing on issues from highway beautification to water quality control, from wilderness preservation to aircraft noise abatement, each study outlines the problem, the nature of proposed legislation, modifications of legislation in the course of congressional decision making, strengths and weaknesses of the final legislative product, and general ability of Congress to respond to the issues at hand.

Cure for Chaos (Fresh Solutions to Social Problems Through the Systems Approach); by Simon Ramo, TRW Inc.; available from David McKay Company, Inc., 750 Third Avenue, New York, N.Y. 10017; 116 pages; 1969.

The systems approach, according to Dr. Ramo, is a cure for a society stricken by too much technology without parallel social advance.   This approach is a methodology, an intellectual discipline, that can logically correlate society's needs and the capabilities of technology.

Decision Making in Air Pollution Control, by George H. Hagevik; published by Praeger Publishers, 111 Fourth Avenue, New York, N.Y. 10003; 232 pages, 1970.

In this volume, the air problem today is defined in terms of the pollutants responsible, the technical aspects of control, the economic issues involved, and the political and institutional framework for dealing with the problem.
Decision strategies that have some utility in air quality management are noted, with the role of bargaining given major emphasis.   Investigations of bargaining behavior are reviewed, and conclusions are derived that are applicable to air quality management.   The point is made that while

bargaining is often viewed as a constraint in the rational decision model, it can also be used to promote efficiency.

Eden in Jeopardy; by Richard G. Lillard; available from Alfred A. Knopf, Publisher, New York, N.Y.; 338 pages; 1967.

This book is a many-sided interpretation of what man has done to Southern California.   In telling that story the author illustrates what men are doing elsewhere in the world and to themselves.

Future Environments of North America; published by The Natural History Press, a division of Doubleday & Company, Inc., 277 Park Avenue, New York, N.Y. 10017; 790 pages; 1967.

Lewis Mumford and 33 other observers of the growing estrangement of man and nature are the contributors to this book.   Describing the transformation of the rural and urban landscapes of America today, the book provides authoritative writings from a wide range of specialties including wildlife management, botany, ecology, geology, economics, city planning, and the law.

Law and Contemporary Problems: Air Pollution Control; is available from the Duke University School of Law, Durham, N.C. 27706; 230 pages; 1968.

The December 1968 issue of *Law and Contemporary Problems* is a symposium of articles on the legal and policy aspects of air pollution control. The symposium's main premise is that the Air Quality Act of 1967 marks a plateau in the development of the air pollution control effort from which it is possible to look both backward and forward and to review the present state of both thinking and action.

The Politics of Pollution, by J. Clarence Davies, III; published by Pegasus Division of Western Publishing Co., Inc., 850 Third Avenue, New York, N.Y. 10022; 232 pages, 1970.

The author describes how government pollution policy is made and analyzes the interests and ideas competing for dominance over pollution control.   His book provides a basis for effective effort and action by both government and the public.

Our Polluted World; by John Perry; published by Franklin Watts, Inc., 575 Lexington Avenue, New York, N.Y. 10022; 214 pages; 1967.

In this book, the origins, nature and effects of air and water pollution are examined in detail, as well as what is being done to alleviate the problems.

The Pollution Reader; compiled by Anthony De Vos, Norman Pearson, P. L. Silveston, and W. R. Drynan; published by Harvest House Ltd., 1364 Greene Street, Montreal, Canada; 264 pages; 1968.

Harvest House of Montreal has published *The Pollution Reader*, based on the "Background Papers" and "Proceedings" of the National Conference on "Pollution and Our Environment" held in Montreal (Oct. 31–Nov. 4, 1966) under the auspices of the Canadian Council of Resource Ministers.

The Smoake of London: Two Prophecies; selected by J. P. Lodge, Jr., National Center for Atmospheric Research, Boulder, Colorado; published by Maxwell Reprint Company, Fairview Park, Elmsford, N.Y. 10523; 56 pages; 1970.

Dr. Lodge has performed a real service by compiling two early writings on air pollution. One, of course, is John Evelyn's seventeenth century work, "Fumifugium: or the Inconvenience of the Aer and Smoake of London Dissipated." This tract, submitted by Evelyn to Charles II, contains a graphic description of the air pollution of London with the suggested remedy of planting "sweet smelling trees." The second selection is a science fiction essay, "The Doom of London." Written by the nineteenth century author Robert Barr, the story is a premonition of the great London smog of 1952.

A Strategy for a Livable Environment; is available from the Superintendent of Documents, U.S. Government Printing Office, Washington, D.C. 20402; 90 pages; 1968.

In November 1966, Secretary John W. Gardner created the Task Force on Environmental Health and Related Problems and charged it with recommending to him goals, priorities, and a Departmental strategy to cope with environmental threats to man's health and welfare.

With Every Breath You Take; by Howard R. Lewis; published by Crown Publishers, Inc., New York, N.Y.; 322 pages; 1965.

Mr. Lewis attempts to dramatize a cause and effect relationship between air pollution and health on a point by point, industry by industry basis, and has produced an outspoken and well-documented book.

B. The Sources and Nature of Air Pollution

Apartment House Incinerators (Flue-Fed), Publication No. 1280; available from the Printing and Publishing Office, National Academy of Sciences, 2101 Constitution Ave., Washington, D.C. 20411; 1965.

The purpose of the study was to obtain answers to problems associated with the design and operation of apartment house incinerators and air pollution.

Atmospheric Emissions from Coal Combustion (PHS Publication No. 999-AP-24) is available from Office of Technical Information and Publications, Environmental Protection Agency, Research Triangle Park, N.C. 27709; 112 pages; 1966.

Information concerning atmospheric emissions arising from the combustion of coal was collected from the published literature and other sources. The data were abstracted, assembled, and converted into common units of expression to facilitate comparison and understanding.

Atmospheric Emissions from the Manufacture of Portland Cement (PHS Publication No. 999-AP-17), by T. E. Kreichelt, D. A. Kemnitz, and S. T. Cuffe; available from Office of Technical Information and Publications, Environmental Protection Agency, Research Triangle Park, N.C. 27709; 47 pages; 1968.

This report summarizes published and unpublished information on actual and potential atmospheric emissions resulting from the manufacture of cement.

Atmospheric Emissions from Nitric Acid Manufacturing Processes (PHS Publication No. 999-AP-27); available from the Manufacturing Chemists Association, Inc., 1825 Connecticut Avenue, N.W., Washington, D.C.; 89 pages; 1967.

Emissions to the atmosphere from the manufacture of nitric acid were investigated jointly by the Manufacturing Chemists Association, Inc. and the U.S. Public Health Service; the study was the second in a cooperative program for evaluation of emissions from selected chemical manufacturing processes.

Atmospheric Emissions from Sulfuric Acid Manufacturing Processes (PHS Publication No. 999-AP-13); available from Manufacturing Chemists

Association, Inc., 1825 Connecticut Avenue, N.W., Washington, D.C.; 127 pages; 1965.

Report, based on joint study conducted by Manufacturing Chemists Association and the U.S. Public Health Service, presents a summary of current air emission control practices in sulfuric acid manufacturing and indicates techniques available for control of all major types of emissions from sulfuric acid plants.

Atmospheric Emissions from Thermal-Process Phosphoric Acid Manufacture (NAPCA Publication No. AP-48); available from Office of Technical Information and Publications, Environmental Protection Agency, Research Triangle Park, N.C. 27709; 74 pages; 1969.

This report, one of a series of cooperative study projects by the Manufacturing Chemists Association and the Public Health Service concerning atmospheric emissions from chemical manufacturing processes, has been prepared to provide information on phosphoric acid manufacture by the wet process.

Atmospheric Emissions from Wet-Process Phosphoric Acid Manufacture (*NAPCA Pub. No. AP-57*); available from Office of Technical Information and Publications, Environmental Protection Agency, Research Triangle Park, N.C. 27709; 96 pages; 1970.

This report, one of a series concerning atmospheric emissions from chemical manufacturing processes, is designed to provide information on phosphoric acid manufactured by the wet process. Background information describing the importance of the wet-process phosphoric acid industry in the United States is included. Basic characteristics of the industry are discussed, including growth rate in recent years, uses for the product, and number and location of producing sites.

Air Pollution Aspects of Tepee Burners (*PHS Publication No. 999-AP-28*); available from the Office of Technical Information and Publications, Environmental Protection Agency, Research Triangle Park, N.C. 27709; 35 pages; 1967.

Based upon the results of a study, the Public Health Service does not consider the use of tepee refuse burners as a suitable method for the disposal of municipal refuse.

Air Pollution Aspects of Brass and Bronze Smelting and Refining Industry (*NAPCA Pub. No. AP-58*); available from Office of Technical Information and Publications, Environmental Protection Agency, Research Triangle Park, N.C. 27709; 64 pages; 1970.

To provide reliable information on the air pollution aspects of the brass and bronze ingot industry, the Brass and Bronze Ingot Institute (BBII) and the National Air Pollution Control Administration, Public Health Service, U.S. Department of Health, Education, and Welfare entered into a cooperative program to study atmospheric emissions from the various industry processes and publish information about them in a form helpful to air pollution control and planning agencies and to brass and bronze industry management.

Carbon Monoxide—A Bibliography with Abstracts (*PHS Publication No. 1503*); available from the Superintendent of Documents, U.S. Government Printing Office, Washington, D.C. 20402; 440 pages; 1967.

Although this annotated bibliography on carbon monoxide includes publications from 1880 to 1966, it merely represents a selection from the wealth of literature on this subject.

Fluid Dynamics of Multiphase Systems, by S. L. Soo, University of Illinois; Blaisdell Publishing Company, a Division of Ginn and Company, 275 Wyman Street, Waltham, Mass. 02154; 524 pages; 1967.

Studies in the dynamics of multiphase systems encompass broad lines of disciplines in engineering and the sciences. The volume covers basic concepts and phenomena of particulate clouds of multiphase systems and cuts across various lines of discipline to regroup the fundamental concepts under a few topics.

Health Aspects of Caster Bean Dust (*PHS Publication No. 999-AP-36*), by E. M. Apen, Jr., W. C. Cooper, R. J. M. Horton, and L. D. Scheel; available from Office of Technical Information and Publications, Environmental Protection Agency, Research Triangle Park, N.C. 27709; 132 pages; 1968.

This publication reviews the occupational and air pollution aspects of castor pomace. An annotated bibliography and selected translations of foreign articles are also included.

Nitrogen Oxides: An Annotated Bibliography (Pub. No. AP-72); available from the Superintendent of Documents, U.S. Government Printing Office, Washington, D.C. 20402; 633 pages; 1970.

The primary objective of this publication is to collect, condense and organize existing literature on the nitrogen oxides. Oxides of nitrogen are important in the air pollution problem, since the complex chain reaction that produces smog is initiated by the photolysis of nitrogen dioxide. In sufficient concentrations, nitrogen oxides themselves may be toxic to humans and vegetation.

Particulate Clouds: Dusts, Smoke and Mists, by H. L. Green and W. R. Lane (Second Edition); published by Spon and D. Van Nostrand Co., Inc., Princeton, N.J.; 482 pages; 1965.

The first edition of this book has become an accepted text in its field, and brought together for the first time in a convenient form all the knowledge available on the subject. This second edition keeps the form of the previous one but incorporates advances which have been made in the science of particles since 1957.

Sulfur Dioxide, Sulfur Trioxide, Sulfuric Acid and Fly Ash: Their Nature and Their Role in Air Pollution (EEI Publication No. 66-900); is available from Edison Electric Institute, 750 Third Avenue, New York, N.Y. 10017; 1218 pages; 1966.

The purpose of this 1200 page monograph is to consider in the context of the air pollution problem the substances stated in the title, and to provide a critical compendium of present knowledge concerning the part played by these substances as air pollutants especially with respect to the health and safety of living organisms.

Sulfur Oxides and Other Sulfur Compounds, (PHS Publication No. 1093); available from the Superintendent of Documents, U.S. Government Printing Office, Washington, D.C. 20402; 396 pages; 1965.

Covering publications from 1893 to 1964, but by no means all-inclusive, this bibliography contains abstracts by subject, title index by subject, author index and geographic location index. Subjects include sources, emission composition and atmospheric reactions; human, animal and plant epidemiology and exposure; control devices and instrumentation development.

Techniques for Controlling Air Pollution from the Operation of Nuclear Facilities; International Atomic Energy Agency; available in the United States from National Agency for International Publications, Inc., 317 East 34th Street, N.Y. 10016; 1966.

The IAEA convened a panel in Vienna, November 4–8, 1963, to correlate present experience in the techniques for controlling nuclear facilities in different countries, to agree on general principles, and to prepare this report.

C. The Transport of Air Pollution

Air Pollution, by Richard Scorer; available from Pergamon Press, Inc., 122 East 55th St., New York, N.Y. 10022; 151 pages; 1968.

The purpose of this impressive little book is, according to the author to describe the basic mechanisms whereby pollution is transported and diffused in the atmosphere, to give practitioners a correct basis for their decisions. It is a book on micrometeorology and it makes the widest possible use of full-color photographs (over 60) to illustrate the behavior of pollution in the atmosphere under a variety of conditions.

Atmospheric Diffusion, by F. Pasquill; published by D. Van Nostrand Company Ltd., 358 Kensington High Street, London W.14, England; 298 pages; 1962.

This book discusses the physical problems which arise from the release of windborne materials into the atmosphere. Proceeding from a review of the analysis of turbulence and diffusion, the author assembles a wealth of data derived from old and recent experiments and provides a study which as a survey of basic research will interest the meteorologist, and as a guide to the assessment of dispersion will meet the needs of other scientists concerned with air pollution.

The Calculation of Atmospheric Dispersion from a Stack; available from Stichting CONCAWE, Gevers Deynootplein 5, The Hague-12, The Netherlands; 70 pages; 1967.

Stichting CONCAWE is a foundation established in the Hague by the Oil Companies' International Study Group for Clean Air and Water Conservation and its name stands for *CON*servation of *C*lean *A*ir and *W*ater, Western *E*urope. The report is the result of a CONCAWE working group review appraising critically the methods of calculating stack height to satisfy a given maximum ground level concentration of pollutant. This has led to a CONCAWE formula for stack height calculation, expressed also as a nomogram.

Descriptive Micrometeorology, by R. E. Munn; available from Academic Press, 111 Fifth Avenue, New York, N.Y. 10003; 245 pages; 1966.

This book fills the need for a comprehensive, balanced, and modern presentation of the continuing research and recent advances in micrometeorology. It deals with the small-scale variations of wind temperature, water vapor, and pollution within 50 meters of the ground. Interactions with the underlying surface and the adjoining planetary boundary layer of the atmosphere are also considered.

The Encyclopedia of Atmospheric Sciences and Astrogeology, edited by Rhodes W. Fairbridge, Columbia University; available from Reinhold Book Division, 430 Park Avenue, New York, N.Y. 10022; 1200 pages; 1968.

Carefully planned and developed as an individual reference source, this comprehensive volume is designed to provide rapid and accurate access to the sources of information that comprise the atmospheric sciences today as seen through the eyes of the earth scientist.

Meteorological Concepts in Air Sanitation, by William P. Lowry and Richard W. Boubel, Oregon State University; available from R. W. Boubel, 3157 Norwood Street, Corvallis, Ore. 97330; 62 pages; 1968.

This book is intended to present to the nonmeteorologist the concepts and language of air pollution meteorology. Necessarily oversimplified and limited in scope, the book will nonetheless enhance any dialogue between the meteorologist and the nonmeteorologist interested in air sanitation problems.

Micrometeorology, by O. G. Sutton; published by McGraw-Hill, Inc., 330 West 42nd Street, New York, N.Y. 10036; 1953.

This classic contains the original contributions by Sir Graham Sutton to the field of micrometeorology.

Recommended Guide for the Prediction of the Dispersion of Airborne Effluents; available from the ASME Order Department, 345 East 47th Street, New York, N.Y. 10017; 85 pages; 1968.

The American Society of Mechanical Engineers has published a guide for predicting the dispersion of airborne pollutants from tall industrial stacks or chimneys.   Guide will enable specialists concerned with air pollution control to estimate the concentration of harmful pollutants at ground level for various stack heights.

Workbook of Atmospheric Dispersion Estimates (PHS Publication No. 999-AP-26); available from Office of Technical Information and Publications, Environmental Protection Agency, Research Triangle Park, N.C. 27709; 1968.

The workbook explains computational techniques currently used by engineers and scientists to solve atmospheric dispersion problems.   Basic working equations are applied to sample problems that exemplify diverse conditions of distance from the source, stack height, wind speed, temperature, stability and related factors.   Also discussed are methods of practical application of dispersion models to estimate concentrations of air pollution.

D. Sampling and Analysis

Air Pollution Monitoring Instrumentation: A Survey (NASA SP-5072); available from Superintendent of Documents, U.S. Government Printing Office, Washington, D.C. 20402; 80 pages; 1969.

The purpose of this survey is to aid in transferring air-pollution-monitoring technology developed in aerospace research to the general industrial domain. Thirty-two instruments and techniques, originally developed for aerospace needs, were covered in this survey.   These were items found to be novel and/or improvements in the area of air pollution monitoring.   A summary of the state of the art in this field precedes the discussion of the above items so that the aerospace-developed items can be compared to current practice.

Air Sampling Instruments (for Evaluation of Atmospheric Contaminants) is available from the American Conference of Governmental Industrial Hygienists, 1014 Broadway, Cincinnati, Ohio 45202; 505 pages; 1969.

This third edition of "Air Sampling Instruments" was proposed by the cooperative efforts of members of the ACGIH Air Sampling Instruments Committee.   As in the first two editions, technical discussions of the principles and usage of air sampling instruments are presented in "A" Sections.   In the "B" Sections the instruments have been classified into groups with a technical discussion of the group followed by individual instrument descriptions.

The Analysis of Air Pollutants, by Wolfgang Leithe; published by Ann Arbor-Humphrey Science Publishers, Inc., Drawer No. 1425, Ann Arbor, Michigan 48106; 304 pages, 1970.

This book described as the first attempt at a complete analysis of problems of air pollution and contamination. It is in itself an original contribution to the solution of those problems connected with air contamination. The book is both an introduction to the study of air analysis and a laboratory handbook, containing detailed instructions of American, British, and German methods of analysis.

An Automatic Sampler for Collecting Radioactive Particulates in Ground Level Air, by D. C. Freeswick, U.S. Atomic Energy Commission; available from Health and Safety Laboratory, U.S. Atomic Energy Commission, 376 Hudson St., New York, N.Y. 10014; 28 pages; 1967.

Since the cessation of nuclear detonations by the United States and Russia; the radioactive contamination of the atmosphere had diluted to the point where it has become necessary to change the analytical techniques and instrumentation for detecting atmospheric radioactive particles.

Basic Gas Chromatography; is available from Varian Aerograph, 2700 Mitchell Drive, Walnut Creek, Calif. 94598; 1967.

The new version of the popular course book, written by Dr. Harold M. McNair and Ernest J. Bonelli, has been expanded and now includes a 21 page laboratory exercise manual. Written in easy to understand language, this practical book can be of tremendous help to the beginning gas chromatographer.

Chemical Detection of Gaseous Pollutants; available from Ann Arbor Science Publishers, Inc., P.O. Box 1425, Ann Arbor, Michigan 48106; 180 pages; 1967.

The olfactory senses are not reliable enough to detect airborne contaminants. A properly used chemical detector can give sufficient information so that qualified persons can appraise intelligently the degree of hazard involved. This abstracted bibliography provides more than 1200 detection for methods 152 gases.

Gas Chromatographic Data Compilation—DS25A; available from American Society for Testing and Materials, 1916 Race Street, Philadelphia, Pa. 19103; 740 pages; 1967.

This compilation of gas chromatography data provides data on the retention indices, column temperature, reference material, liquid phase, and solid support for some 3800 compounds. Retention indices have been added to provide useful data for the theoretician.

The Particle Atlas, by McCrone, Draftz, and Delley; published by Ann Arbor Science Publishers, Inc., P.O. Box 1425, Ann Arbor, Michigan 48106; 424 pages; 1967.

This reference volume is a complete guide to the positive identification of fine particles, dusts, and contaminants. Based on the concept of comparing full-color particle photomicrographs with the visual image seen through a microscope, the book is accompanied by the claim that any detectable particle can be identified. With 512 full-color photomicrographs, each one $2\frac{5}{8} \times 3\frac{3}{8}$ in., plus a binary classification system extending the range of identification, the Atlas seems to substantiate this claim.

Rapid Survey Technique for Estimating Community Air Pollution Emissions (*Public Health Service Pub. No. 999-AP-29*); available from Office of Technical Information and Publications, Environmental Protection Agency, Research Triangle Park, N.C. 27709; 77 pages; 1967.

A method is presented for estimating rapidly the major emissions of air pollutants in a community. The method is based on information that is readily available in most urban areas; it does not entail extensive surveys or sampling procedures.

Quantitative Analysis of Gaseous Pollutants, by Walter E. Ruch, Ph.D.; published by Ann Arbor-Humphrey Science Publishers, Inc., Drawer No. 1425, Ann Arbor, Michigan 48106; 241 pages; 1970.

Dr. Ruch presents 376 methods for the quantitative analysis of gaseous pollutants—methods abstracted with thoroughness and consistency to enable one to quickly find the method best suited to his needs. Each abstract contains a cross reference to related abstracts, a basic outline of the method with specific quantities and procedure, a statement regarding the concentration range for which the method is applicable, the sampling procedure and equipment used in detail, interferences which could alter the determination, a closing statement regarding the suitability of the method for industrial hygiene or air pollution work and the time required to complete the analysis.

Selected Methods for the Measurement of Air Pollutants (PHS Publication No. 999-AP-11); available from Office of Technical Information and Publications, Environmental Protection Agency, Research Triangle Park, N.C. 27709; 54 pages; 1965.

This manual is an effort to assist in the development of uniform standard methods of analysis of air pollutants. It makes available the judgment and knowledge of a large group of chemists in the Public Health Service. Methods of determining pollutants of common interest are presented in uniform format by chemists on the staff of the Division of Air Pollution.

Standards and Practices for Instrumentation, edited by James E. French; available from Instrument Society of America, Publications Department, 530 William Penn Place, Pittsburgh, Pa. 15219; 1967.

This 550 page second edition is reported to be the most complete single source for instrumentation standards and practices ever compiled. In addition to 33 complete ISA Recommended Practices, the book includes abstracts of over 500 instrumentation and control standards developed by 24 technical societies and the British Standards Institution, Canadian Standards Association, International Electrotechnical Commission, and the International Organization for Standardization.

E. Economic Effects

Economic Costs of Air Pollution: Studies in Measurement, by Ronald G. Ridker; published by Frederick A. Praeger, Publishers, 111 Fourth Avenue, New York, N.Y. 10003; 228 pages; 1967.

In this pioneer study, Ronald G. Ridker establishes methods to estimate the cost of air pollution, a cost which includes health and aesthetic losses as well as losses to property. While estimates of actual cost are provided wherever possible, this work is primarily meant to determine what can and what needs to be done to measure the economic cost of air pollution.

The Economics of Air Pollution, ed. by H. Wolozin; published by W. W. Norton & Co., Inc., 55 Fifth Avenue, New York, N.Y. 10003; 318 pages; 1966.

What has air pollution cost us as a nation and what will it cost us to decontaminate the air? Who will pay for the air pollution control programs? How can the price of the control be allocated? Should industry bear the cost alone? The chapters in this book systematically describe and dissect economic theories and techniques as they presently apply, and as they may be applied, to air contamination.

Environmental Quality in a Growing Economy, edited by Henry Jarrett; published for Resources for the Future, Inc., by The Johns Hopkins Press, Baltimore, Maryland 21218; 173 pages; 1968.

The Resources for the Future 1966 Forum brought together a dozen scholars who explored the nature and dimensions of this quality problem largely from the standpoint of economics and the other social sciences. Their papers, as revised by the authors for publication, constitute the present volume.

Recognition of Air Pollution Injury to Vegetation: A Pictorial Atlas; available from the Air Pollution Control Association, 4400 Fifth Ave., Pittsburgh, Pa 15213; 112 pages; 1970.

A cooperative effort of the TR-7 Agricultural Committee of the Air Pollution Control Association and the National Air Pollution Control Administration, this book describes and illustrates, with 120 color plates, the visible effects of pollutants on vegetation. It contains an assessment of these effects on the total environment, and includes tables of relative sensitivity.

F. Health Effects

Air Quality Criteria for Particulate Matter (No. AP-49); published by the Office of Technical Information and Publications, Environmental Protection Agency, Research Triangle Park, N.C. 27709; 211 pages; 1969.

Air quality criteria tell us what science has thus far been able to measure of the obvious as well as the insidious effects of air pollution on man and his environment. Such criteria provide the most realistic basis that we presently have for determining to what point the levels of pollution must be reduced if we are to protect the public health and welfare. This document, first in a continuing series, focuses its attention on atmospheric particulate matter.

Air Quality Criteria for Sulfur Oxides (No. AP-50); published by the Office of Technical Information and Publications, Environmental Protection Agency, Research Triangle Park, N.C. 27709; 178 pages; 1969.

Air Quality Criteria for Carbon Monoxide (Pub. No. AP-62); available from the Office of Technical Information and Publications, Environmental Protection Agency, Research Triangle Park, N.C. 27709; 1970.

Air Quality Criteria for Photochemical Oxidants (Pub. No. AP-63); available from the Office of Technical Information and Publications, Environmental Protection Agency, Research Triangle Park, N.C. 27709; 1970.

Air Quality Criteria for Hydrocarbons (Pub. No. AP-64); available from the Office of Technical Information and Publications, Environmental Protection Agency, Research Triangle Park, N.C. 27709; 1970.

Inhaled Particles and Vapours II; Pergamon Press Inc., 44–01 21st Street, Long Island City, N.Y. 11101, 605 pages; 1967.

This volume presents the proceedings of the Second International Symposium organized by the British Occupational Hygiene Society, at Cambridge. Edited by C. N. Davies, the book contains 48 papers on topics including the mechanics of breathing, fine anatomy of the lung, deposition of particles in various parts of the lung, the uptake of vapors, the elimination and storage of various dusts in the lungs.

Seminar on Human Biometeorology (PHS Publication No. 999-AP-25); available from Office of Technical Information and Publications, Environmental Protection Agency, Research Triangle Park, N.C. 27709; 183 pages; 1968.

This volume is a collection of papers presented at Cincinnati, Ohio, Jan. 14–17, 1964, at a seminar on human biometeorology. Topics discussed included physiological and climatological instrumentation, climates of the United States, altitude, microclimatology, indoor and outdoor weather, ultraviolet light, heat exposure, air ions, and cold stress.

Toxicologic and Epidemiologic Bases for Air Quality Criteria; a reprint from the Journal of the Air Pollution Control Association, Pittsburgh, Pa.; 104 pages; 1969.

This special report consists of papers, discussion, and comment from three sessions dealing with air quality criteria at the 62nd Annual Meeting of the Air Pollution Control Association. The format of this special report comprises Part 1: particulate matter, oxides of sulfur, and sulfuric acid; Part 2: oxidants, hydrocarbons, and oxides of nitrogen; and Part 3: carbon monoxide and lead. In each part, air quality criteria are discussed from three standpoints: environmental appraisal, toxicologic appraisal, and epidemiologic appraisal. Each principal paper is followed by a discussion.

G. The Control of Air Pollution

Acoustic Coagulation and Precipitation of Aerosols; by Eugenii Pavlovich Mednikov; Authorized translation from the Russian by Chas. V. Larrick; published by Consultants Bureau Enterprises Inc., 227 West 17th Street, New York, N.Y. 10011; 180 pages; 1965.

This is of exceedingly great interest in dust and drop collecting technology, which occupies an important place in present day industry. Preliminary acoustic coagulation of such particles makes precipitation by dust and drop collecting equipment substantially more efficient. In addition, according to this report, sound waves not only make the suspended particles easier to coagulate, but can be applied directly to the precipitation process.

Air Pollution Engineering Manual (PHS Publication No. 999-AP-40); available from the Office of Technical Information and Publications, Environmental Protection Agency, Research Triangle Park, N.C. 27709; 892 pages; 1968.

This manual discusses in detail numerous design problems confronting air pollution engineers in the development of control systems, and describes control techniques and equipment for such sources as metallurgical and mechanical processes, processes of incineration and combustion, and processes associated with petroleum and chemical equipment. It was developed by the staff of the Los Angeles County Air Pollution Control District.

Air Pollution Manual: Part 2—Control Equipment; published by the American Industrial Hygiene Association, 14125 Prevost, Detroit, Michigan 48227; 150 pages; 1968.

This manual is the second in a two-volume set published by the American Industrial Hygiene Association on the subject of air pollution. Volume I covered the general aspects of air pollution, its effects, atmospheric sampling, and administration of control programs. The new volume is concerned with the equipment for control of air pollution sources and the proper application of such equipment. It is written for the practicing plant or process engineer.

A Compilation of Selected Air Pollution Emission Control Regulations and Ordinances; is available from the Office of Technical Information and Publications, Environmental Protection Agency, Research Triangle Park, N.C. 27709; 1966.

The regulations and ordinances have been arranged in such a manner that each section of this report is a compilation of laws pertaining to a specific type of pollutant or pollutant source. The regulations and ordinances compiled were selected to represent the different methods of controlling emissions by law and to represent varying degrees of control.

Control of Airborne Dust; by W. D. Bamford; published by the British Cast Iron Research Association, Alvechurch, Birmingham, England; 480 pages; 1961.

Although the work described in this book was primarily directed to the conditions prevailing in iron foundries, it should prove of great interest and value in any industry where the control of dust, fume and air flow has to be considered.

Control of Air Pollution; by Alan Gilpin; published by Butterworth & Co. Ltd., London, England; 514 pages; 1963.

This book was written by a local government officer for other local government officers as well as for industrial users of fuel and energy and candidates for the Public Health Inspector's Diploma. The author discusses different types of furnaces, fuels and firing methods, and studies in detail the practical application of the Clean Air Act of 1956.

Control Techniques for Particulate Air Pollutants (No. AP-51); published by the Office of Technical Information and Publications, Environmental Protection Agency, Research Triangle Park, N.C. 27709; 216 pages; 1969.

The control techniques described herein represent a broad spectrum of information from many engineering and other technical fields. The devices, methods, and principles have been developed and used over many years, and much experience has been gained in their application. They are recommended as the techniques generally applicable to the broad range of particulate emission control problems.

Control Techniques for Sulfur Oxide Air Pollutants (No. AP-52); published by the Office of Technical Information and Publications, Environmental Protection Agency, Research Triangle Park, N.C. 27709; 122 pages; 1969.

This is the second of a continuing series of documents to be produced under the federal program established to develop and distribute control technology information. This document is the result of intensive effort on the part of many people, and represents the techniques presently applicable for controlling sulfur oxide air pollution.

Control Techniques for Carbon Monoxide Emissions from Stationary Sources (Pub. No. AP-65); available from the Office of Technical Information and Publications, Environmental Protection Agency, Research Triangle Park, N.C. 27709; 1970.

Control Techniques for Carbon Monoxide, Nitrogen Oxides and Hydrocarbon Emissions from Mobile Sources (Pub. No. AP-66); available from the Office of Technical Information and Publications, Environmental Protection Agency, Research Triangle Park, N.C. 27709; 1970.

Control Techniques for Nitrogen Oxides Emissions from Stationary Sources (Pub. No. AP-67); available from the Office of Technical Information and Publications, Environmental Protection Agency, Research Triangle Park, N.C. 27709; 1970.

Control Techniques for Hydrocarbons and Organic Solvent Emissions from Stationary Sources (Pub. No. AP-68); available from the Office of Technical Information and Publications, Environmental Protection Agency, Research Triangle Park, N.C. 27709; 1970.

Cupola Emission Control; is available from Gray and Ductile Iron Founders' Society, National City, East Sixth Building, Cleveland, Ohio 44114; 196 pages; 1969.

It is an edited translation of *Kupolofenententstaubung* written by Gerhard Engels and Ekkehard Weber of Verein Deutscher Giessereifachleute, the association of German Foundrymen. The principal purpose of this book is to provide engineering information on the nature and characteristics of ferrous cupola emissions and the factors that must be considered in their control techniques.

Gas Purification Processes; edited by G. Nonhebel; available from The Chemical Rubber Company, 2310 Superior Avenue, Cleveland, Ohio 44114; 894 pages; 1966.

Prepared by 29 worldwide specialists, this book brings together the wide range of information on the many processes of gas purification employed in chemical works. Including the latest data on air pollution control, its ambitious scope encompasses general principles to practical details.

Industrial Electrostatic Precipitation; by Harry J. White; published by Addison-Wesley Publishing Company, Inc., Reading, Mass.; 376 pages; 1963.

This book provides the first comprehensive account of the basic principles of electrostatic precipitation, the most widely used process for the control of air pollution in heavy industry. It emphasizes the science and technology essential to the many important applications of this process. The material presented was gathered over a 20 year period of research, development, and field experience by the author.

Industrial Gas Cleaning: The Principles and Practice of the Control of Gaseous and Particulate Emissions; by W. Strauss, Reader in Industrial Science, University of Melbourne; published by Pergamon Press Inc., 44-01 21st Street, Long Island City, N.Y. 11101; 472 pages; 1966.

Volume 8 in the International Series of Monographs in Chemical Engineering is an integrated account of the collection of particles and gases, with particular reference to the control of air pollution. It relates the basic mechanisms of particle collection to their application in the design of gas cleaning plant, cyclones, filters, electrostatic precipitators, scrubbers, settling chambers, and gas absorption systems.

Industrial Waste Disposal; edited by Richard D. Ross, Thermal Research & Engineering Corp.; published by Reinhold Book Division, 430 Park Avenue, New York, N.Y. 10022; 340 pages; 1968.

The book can be of great help to anyone who must cope with the problem of industrial waste disposal. It provides basic facts and ideas that can be used in formulating solutions to future problems, and these, in turn, will lead to a fuller understanding of the operations and technologies themselves.

Pollution Control (Chemical Process Monograph No. 10); published by
Noyes Development Corporation, 16–18 Railroad Avenue, Pearl River,
N.Y. 10965; 207 pages; 1965.

This technical/economic study attempts to discuss proved and proposed
schemes for solving the problems of pollution control.

Power Systems for Electric Vehicles (PHS Publication No. 999-AP-37);
available from the Office of Technical Information and Publications,
Environmental Protection Agency, Research Triangle Park, N.C. 27709;
320 pages; 1968.

How far current industrial, academic, and governmental research has
progressed toward development of commercially practical power systems for
electric vehicles is summarized in this publication.

IV. PERIODICAL LITERATURE

A. Journals and Magazines

1. Australia

Clean Air, Journal of the Clean Air Society of Australia & New Zealand;
Department of Industrial Science; University of Melbourne; Parkville,
Victoria 3052, Australia.

This is the official publication of the Clean Air Society of Australia and
New Zealand.   It is directed towards those whose aim is to conserve clean
air in Australia and New Zealand.

2. England

Analyst; The Journal of the Society for Analytical Chemistry; W. Heffer
and Sons, Ltd.; Cambridge, England.

The Official journal of the Society of Analytical Chemists is an international
publication directed towards all branches of analytical chemistry.

Atmospheric Environment (see United States).

Journal of the Institute of Fuel; 18 Devonshire St.; Portland Place; London
W1, England.

This journal is aimed at engineers and research men in the fields of fuel
combustion and related subjects.

Journal of Scientific Instruments; The Institute of Physics and The Physical Society; 1 Lowther Gardens; Prince Consort Road; London SW7, England.

Aimed at scientists involved in the development and use of scientific instruments.

Nature; Macmillan and Co., Ltd.; 4 Little Essex St.; London WC2, England.

Nature is an international science journal directed towards scientists working in the fields of physical and biological sciences.

Quarterly Journal of the Royal Meteorological Society; 49 Cromwell Road.; London SW7, England.

The primary purpose of the Quarterly Journal is to publish results of original research and reports of the Society's scientific meetings.

Clean Air (formerly *Smokeless Air*); National Society for Clean Air; 134/137 North Street, Brighton BN1 1RG, England.

Published by the National Society for Clean Air this quarterly journal deals largely with news of the field with an occasional article aimed at the clean air protagonist.

Environmental Pollution; Elsevier Publishing Co. Ltd.; Ripple Road, Barking, Essex, England.

A journal concerned principally with the biological and ecological effects of all types of pollution and pollution control.

3. France

Pollution Atmosphérique; 22 rue Murillo; 75 Paris (8e), France.

This journal deals with all phases of air pollution.   It is directed towards the French scientist and engineer dealing with air pollution problems.

Revue Generale de Thermique; 2 rue des Tanneries; Paris (13e), France.

This journal covers the field of fuels, energy, and thermal equipment. It is the publication resulting from the merger of the journals "Chaleur et Industrie" and "Flame et Thermique."

4. Germany

Brennstoff-Wärme-Kraft; VDI Verlag GmbH; Düsseldorf, Germany.

The science and technology of energy production are the principal subjects of this journal. It is directed towards the German scientist and engineer.

Staub; VDI-Fachgruppe Staubtechnik; Bongardstr. 3; Düsseldorf, Germany.

This journal is published for the air pollution scientist interested in all phases of the problem.

Chemie-Ingenieur-Technik; Verlag Chemie GmbH; Pappelallee 3, 694 Weinheim, Germany.

This publication is directed toward chemical engineers and other technical workers in the chemical process industry.

5. Italy

Fumi & Polveri; Via Soperga 52; Milan, Italy.

This journal is intended for the use of scientists and engineers engaged in air pollution control work.

6. Japan

Netsu-Kanri [Heat Management]; Chuo Netsu Kanri Kyogikai 8, Ichigaya-kawada-cho; Shinjuku-ku; Tokyo, Japan.

This is the English translation of the official publication of the Central Association for Heat Management (Japan). Its subtitle and specific area of interest is "Energy and Pollution Control."

7. United States

American Industrial Hygiene Assoc. Journal; Department of Industrial Health; School of Public Health; The University of Michigan; Ann Arbor, Michigan 48104.

Distributed to members of the Association. Technical and scientific reports relating to industrial hygiene, occupational health.

Analytical Chemistry; American Chemical Society; 1155 16th St., N.W.; Washington, D.C. 20006.

Research engineers, chemists, physicists, and metallurgists. Detailed theoretical and applied reports dealing with instrument applications, research equipment and apparatus, materials, chemicals, etc.

Archives of Environmental Health; American Medical Assoc.; 535 N. Dearborn St.; Chicago, Ill. 60610.

Official publication of the Amer. Academy of Occupational Medicine. Editorial content for physicians specializing in industrial hygiene.

Atmospheric Environment; Pergamon Press; 44-01 21st St.; Long Island City, N.Y. 11101.

An international journal covering subjects on all aspects of air pollution research which are of other than purely local interest.

CRC Critical Reviews in Environmental Control; The Chemical Rubber Company; 18901 Cranwood Parkway; Cleveland, Ohio 44128.

Quarterly journal contains at least four articles extracting the most significant papers from the mass of current literature, focusing attention on the major developments in environmental control.

Chemical Engineering; 330 West 42nd St.; New York, N.Y. 10036.

Edited for chemical engineers and other technical decision makers in the chemical process industry who develop, design, build and operate plants that utilize chemical technology.

Environmental Science & Technology; American Chemical Society; 1155 Sixteenth St., N.W.; Washington, D.C. 20036.

An ACS publication. It serves the pollution control market: water pollution, air pollution, and waste treatment. Edited for engineers, scientists, and officials in industry, government and university pollution control areas.

Foundry; Penton Publishing Company; Penton Bldg.; 1213 West Third St.; Cleveland, Ohio 44113.

Edited for management, production, engineering, research, and technical personnel in foundries and metal casting departments.

Health Physics; Pergamon Press, Inc.; Maxwell House; Fairview Park; Elmsford, N.Y. 10523.

This journal is published in Ireland. It is the Official Journal of the American Health Physics Society. It is directed towards an international audience of health physics scientists with emphasis on radiation protection, radiation physics, radiation dosimetry, radioecology, etc.

Hygiene and Sanitation; Engl. transl. of Gigiena i Sanitariya; Clearinghouse for Federal Scientific and Technical Information; U.S. Dept. of Commerce; Springfield, Va. 22151.

The English Translation of Gigiena i Sanitariya is directed towards hygienists and sanitation physicians.

Journal of the Air Pollution Control Assoc.; Air Pollution Control Assoc.; 4400 Fifth Ave.; Pittsburgh, Pa. 15213.

Official publication of the Air Pollution Control Association. It is edited for those concerned with an understanding of and solutions to problems of air pollution, including control agencies, industry, research, educational, and consulting groups.

Journal of Applied Meteorology; American Meteorological Society; 45 Beacon St.; Boston, Mass. 02108.

This journal is published by the American Meteorological Society as a medium for articles of an applied nature.

Journal of Atmospheric Sciences; American Meteorological Society; 45 Beacon St.; Boston, Mass. 02108.

The Journal of the Atmospheric Sciences is a medium for the publication of research related to the atmospheres of the earth and other planets. It is published bimonthly by the American Meteorological Society. The editorial policy is to encourage papers which emphasize the quantitative and deductive aspects of the subject.

Journal of Geophysical Research; American Geophysical Union; 1515 Massachusetts Ave., N.W.; Washington, D.C. 20005.

The official publication of the American Geophysical Union, it publishes original scientific contributions on the physics and chemistry of the earth, its environment, and the solar system.

Journal of Physical Chemistry; American Chemical Society; 1155 Sixteenth St., N.W.; Washington, D.C. 20036.

Edited for chemical physicists and physical chemists. Articles deal with fundamental concepts, atomic and molecular phenomena, and systems for which clearly defined models are applicable.

Law and Contemporary Problems; Duke Station; Durham, N.C. 27706.

Edited by the Duke University School of Law, this journal examines contemporary problems in depth.

Mechanical Engineering; American Society of Mechanical Engineers; 345 East 47th St.; New York, N.Y. 10017.

Edited for graduate engineers involved in: original equipment design and production; energy transformation and transmission; plant and process engineering.

Modern Manufacturing; (formerly *Factory*); McGraw-Hill, Inc.; 330 West 42nd St.; New York, N.Y. 10036.

Intended for men in production engineering, maintenance, and plant management. It provides basic technical and operational help in manufacturing.

Oil and Gas Journal; The Petroleum Publishing Co.; 211 S. Cheyenne Ave.; Tulsa, Okla. 74101.

Edited for operating and management personnel of worldwide oil and gas operations.

Phytopathology; The American Phytopathological Society; 1821 University Ave.; St. Paul, Minn. 55104.

The official publication of the American Phytopathological Society. It publishes research articles and review articles on plant pathology.

Product Engineering; McGraw-Hill, Inc.; 330 West 42nd St.; New York, N.Y. 10036.

A technical news magazine for design engineers responsible for design and development of products for resale or in-plant equipment.

Society of Automotive Engineers Journal; 485 Lexington Ave.; New York, N.Y. 10017.

The official organ of the society; its aim is to keep automotive engineers currently aware of technical progress made throughout the industry.

Science; American Assoc. for the Advancement of Science; 1515 Massachusetts Ave., N.W.; Washington, D.C. 20005.

Published by the American Association for the Advancement of Science, this reaches a wide range of scientists. It provides a forum for the presentation and discussion of important issues related to the advancement of science, including the presentation of minority or conflicting points of view.

Staub (Engl. transl.); The Clearinghouse for Federal Scientific and Technical Information; U.S. Department of Commerce; Springfield, Va. 22151.

The English translation of the German publication is directed towards those interested in air pollution control problems.

TAPPI (Technical Assoc. of the Pulp & Paper Industry); 360 Lexington Ave.; New York, N.Y. 10017.

The official publication of the Technical Association of the Pulp and Paper Industry is edited for operating executives, engineers, scientists, plant supervisors and technical advisors in pulp, paper, paperboard, and converting industries. Articles report on process design and development, product research and development, original research, and management practices.

B. Newsletters

Air Pollution Advisory Newsletter; College of Agriculture and Environmental Science; Rutgers University—The State University of New Jersey; New Brunswick, N.J., 08903.

Air and Water Conservation News; American Petroleum Institute; 1271 Avenue of the Americas; New York, N.Y. 10020.

Air and Water News (Weekly); McGraw-Hill, Inc.; 330 West 42nd Street; New York, N.Y. 10036.

Air/Water Pollution Report (Weekly); Business Publishers, Inc.; P.O. Box 1067, Blair Station; Silver Spring, Maryland 20910.

Air/Water Research Briefs; American Petroleum Institute; Committee on Public Affairs; 1271 Avenue of the Americas; New York, N.Y. 10020.

Clean Air and Water News (Weekly); Commerce Clearing House, Inc.; 4025 W. Peterson Avenue; Chicago, Ill. 60646.

Conservation Foundation Letter; The Conservation Foundation; 1250 Connecticut Avenue, N.W.; Washington, D.C. 20036.

Currents; Manufacturing Chemists Association; 1825 Connecticut Avenue, N.W.; Washington, D.C. 20009.

Environmental Health Letter (Bimonthly); Gershon W. Fishbein, Publisher; National Press Building; Washington, D.C. 20004.

Environmental Spectrum Newsletter; College of Agriculture and Environmental Science; Rutgers University—The State University of New Jersey; New Brunswick, N.J. 08903.

Environmental Technology and Economics; Technomic Publishing Co., Inc.; Stamford, Connecticut.

Environment Reporter; The Bureau of National Affairs, Inc.; 1231 25th Street, Northwest; Washington, D.C. 20037.

Monthly Bulletin; National Council of the Paper Industry for Air and Stream Improvement; New York, N.Y.

National Air Conservation Commission Newsletter; National Tuberculosis and Respiratory Disease Association; 1740 Broadway; New York, N.Y. 10019.

Response; Office of Information, U.S. Department of Agriculture; Washington, D.C. 20250.

Waste Management Report; Patton-Clellan Publishing Co.; Washington, D.C.

Water–Air Legislative Report; American Paper Institute; New York, N.Y.

AUTHOR INDEX

SUBJECT INDEX